Marc Achtelig

**"Technical Documentation Best Practices" Series:
Creating Effective Visualizations for Technical Communication**

Images, Videos, Interactive Content

First Edition

indoition

indoition publishing e.K.
Goethestr. 24
90513 Zirndorf near Nürnberg, Germany

Tel.: *+49 (0)911/60046-659*
Fax: *+49 (0)911/60046-863*
Email: *info@indoition.com*
Internet: *www.indoition.com*

Author: Marc Achtelig
Design: Marc Achtelig
Cover picture: iStock.com/procurator

Trademarks
All terms mentioned in this book known by the publisher and by the author to be trademarks or service marks have been appropriately capitalized. However, the publisher and the author cannot attest to the accuracy of this information. Use of a term in this book should not be regarded as information whether a trademark or service mark does exist or does not exist.

Warnings and Disclaimer
Every possible effort has been made to ensure that the information contained in this book is accurate and complete, but the publisher and the author cannot accept responsibility for any errors or omissions, however caused. No warranty or fitness is implied. The publisher and the author shall have neither liability nor responsibility to any person or entity with respect to any loss or damage arising from the information contained in this book or from the lack of information not contained in this book. The book doesn't provide any individual advice; in particular, it doesn't provide any legal advice. Before shipping your products and before publishing any content, make sure that you follow all relevant standards, laws, and other regulations that are applicable for both your own country and for all countries to which you will sell and ship your products. All rules that are given in these standards, laws, and other regulations take precedence over the recommendations given in this book.

ISBN 978-3-943860-10-8

© Copyright 2020 by indoition publishing e.K.
First Edition

This work is protected by copyright. Its use outside of the restricted limits of the copyright law is not permitted without the written, legally valid approval of the publisher and is a punishable offense. This particularly applies to the copying, translating, microfilming, as well as saving and processing in electronic systems. All rights reserved.

For permissions to reproduce selections from this book, please write to *info@indoition.com* or to the postal address or fax number given above.

About the author

Marc Achtelig has been in the technical communication business since 1989.

After some years in the development of simulation and educational software, Marc joined one of the major technical communication services providers as technical writer, information architect, and consultant. In 2004, he founded his own consulting business.

Marc was one of the pioneers and early evangelists of single source publishing, the approach to create printed manuals and online help systems from the same shared text base. He has published many articles and several books and has spoken at various national and international conferences.

He holds degrees in process engineering and industrial engineering.

For individual consulting services and trainings, contact Marc at: *ma@indoition.com*.

Contents

1	**Preface**	9
2	**Visualizing**	11
2.1	**Which medium to use?**	13
2.2	**Common basics of visualization**	19
	Avoid visual noise	20
	Use color sparingly	25
	Use symbols sparingly	28
	Include text elements when necessary	33
	Tips on writing text within visuals	37
	Carefully consider whether to show persons	40
	Be aware of cultural differences	41
	Avoid legal pitfalls	44
2.3	**Images in general**	47
	Use each image for a clear purpose	49
	Position each image precisely	51
	Page layout of pages with images	55
	Balance the visual weight of images with headings	60
	When to add borders around images?	62
	What image size?	64
	Indicate the size and context of the subject	67
	Guide the users' eyes	70
	Emphasize what's important	77
	Zoom in on details	79
	Use text callouts or a legend for labeling?	82
	Tips on formatting callouts and legends	85
	Tips on formatting image texts in general	95
	You rarely need figure titles	101
	You rarely need figure numbers	104
	Color or black and white?	107

Which resolution? ... 111
Which color mode? ... 116
Which file format? ... 118
Think ahead about editing an image ... 121
Think ahead about reusing an image ... 123
Think ahead about translating the texts in an image ... 125
Tips for developing graphics on your own ... 128
Tips for working with graphic artists and photographers ... 131

2.4 Images of hardware ... 135
Photo or drawing? ... 136
Which perspective? ... 138
Ways of showing the interior of objects ... 143
Ways of indicating dynamics ... 146
Prepare a photo session with care ... 150
Tips for taking photos of technical devices ... 152
Tips for editing photos of technical devices ... 164

2.5 Images of software ... 167
How many screenshots to show? ... 169
Use a real screenshot or an illustration? ... 174
What to show in a screenshot? ... 177
When to include the mouse pointer? ... 183
Don't maximize the windows to capture ... 186
Get rid of dead space ... 189
Avoid confusion with the real UI ... 193
Use standard settings ... 194
With web applications, don't show the browser ... 197
Show meaningful data ... 199
Hide private data ... 202
If it saves you time, fake a screenshot ... 205
Optimize each screenshot for its particular purpose ... 207
Tips for taking screenshots ... 209
Tips for translating screenshots ... 214

2.6 Video design ... 219
Which type of video? ... 220

	Keep videos short	224
	Provide navigation and orientation	226
	Where to place the videos?	230
	Use meaningful poster images and titles	233
	Create a storyboard	235
	Add text, closed captions, or voice-over?	240
	How to handle warnings?	243
	Show the presenter?	245
	One or two presenters?	248
	Male or female presenters?	250
	Professional voice, own voice, or synthetic voice?	251
	Use music?	253
	Include ambient noise?	255
	Keep the video simple	257
	Use effects wisely	262
	Consider embedding also still images	267
	Consider linking to further information	268
	Consider adding a call to action	270
	Consider creating a brand	272
	Create small video modules	274
	Standardize your video modules	276
2.7	**Video production**	**279**
	Rethink you idea of "good"	280
	Prepare each shooting with care	282
	Clothing	284
	Lighting	286
	Tips for recording video and audio	292
	Tips for presenting yourself on camera	296
	Tips for recording your own voice	298
	Tips for localizing and translating your videos	301
	Tips for keeping your videos up to date	303
	In your videos, take your time	305
	Consider keeping mistakes visible	307
2.8	**Interactive content**	**309**
	Interactive 2D images	310
	Interactive 3D images	313

7

Interactive video (hypervideo) 316
Augmented reality 318
Scripts and more 322

3   References and further information 327

1 Preface

Welcome to this book, your companion to illustrating and animating user assistance and to creating instructional videos.

What you will find in this book

This book provides you with effective rules and tips that will help you to create visuals, videos, and interactive content that communicate technical information clearly. It starts with the general principles that apply to both images and videos and then details on the different types of visual content in particular:

- images of hardware: photos and drawings
- images of software: screenshots
- videos
- interactive content, such as interactive images, interactive videos (hypervideos), augmented reality applications, and more

Each topic begins on a new page, so skimming through the book is easy. You don't have to read everything from start to finish. All topics are independent of each other. To be able to understand a particular topic, you don't need to have read certain other topics before.

What you won't find in this book

The book provides clear rules and unambiguous recommendations. No boring theory, no musings, no shoptalk.

However, because each product and audience is different, there are often no ready-made, one-size-fits-all solutions. The book introduces you to the basic principles, shows you what's important, and can inspire you. However, the book cannot make any individual decisions for you.

Please consider the presented rules as general recommendations, not as laws that must be followed slavishly.

> **ⓘ Important:** The book can't provide individual advice; in particular, it can't provide any legal advice. Before shipping your products and before publishing any content, make sure that you also follow all relevant standards, laws, and other regulations that are applicable for both your own country and for all countries in which you sell your products. All rules that are given in these standards, laws, and other regulations take precedence over the recommendations given in this book.

This book is about perfection—but it's not perfect itself

When reading this book, you may notice that it sometimes doesn't manage to follow its own advice. You may find typos, grammar mistakes, and things that could have been said and depicted more clearly. Also, there are always some technical limitations. Ouch—sorry!

Everyone involved in creating this book took every effort not to overlook anything and to make every detail perfect. We've used some of the best spelling checkers, grammar checkers, and writing enhancement software. We had the text double-checked by human editors. Yet, there are still some mistakes. On closer inspection, no book is perfect. This one isn't either.

We hope that this book will provide enough value so that you will forgive us for its own imperfections.

Have a good time reading ;-)

2 Visualizing

A picture can say more than a thousand words, and a video can say even more than a thousand pictures. Well, really?

It depends ....

To do an effective job, images and videos need to be used where they really make sense, and they each need to be designed in a way that communicates their particular message clearly.

This book teaches you the principles of effective visualization:

- *Which medium to use?* 13
  Helps you decide whether to use text, an image, a video, or interactive content to communicate some particular information.

- *Common basics of visualization* 19
  Summarizes the basic things that you need to keep in mind for creating both images and videos.

- *Images in general* 47
  Provides tips for creating effective images in general.

- *Images of hardware* 135
  Provides tips for creating images that show physical devices. These images can be both drawings and photos.

- *Images of software* 167
  Provides tips for creating images that show software (screenshots).

- *Video design* 219
  Provides tips for designing effective instructional videos and on how to best embed these videos into technical documentation.

- *Video production* 279
  Focuses on the technical and organizational aspects of producing instructional videos.

- *Interactive content* 309
  Gives you ideas on implementing interactive components, such as interactive 2D and 3D images, hypervideo, and augmented reality applications and more.

2.1 Which medium to use?

Effective documents use both verbal and visual coding. Some information can best be given by words, but other information can better be given by still images or by moving images (video).

Instructions that use only text and no images, or instructions that use only images but no text, are often hard to understand.

Each medium (text, images, videos) has its particular strengths and weaknesses. To make your documents as user-friendly and effective as possible, use the different media flexibly:

- Where text conveys your message easily, use **text**.
- Where a still image conveys your message more easily than text, use an **image**.

  However, never use an image only for decoration or because you think that it "looks better" than text.
- In online documentation: Where moving images (videos, screencasts) convey your message more easily than text or still images, use a **video**.

  However, never use video to impress.
- In online documentation: Where it helps users to engage and experiment, even consider adding some **interactive elements** rather than a prefabricated video.

Try a little thought experiment:

Attempt to show to a group of aliens why we need to eat some food—or simply explain it with text ....

Attempt to explain to the aliens with words how to identify an apple—or simply show them a picture of an apple ....

Attempt to explain to the aliens with words how to peel the apple—or simply show them a short video ....

Pros and cons of text

Pros:

- Text may be legally required to have in some countries and for some products.
- Text can be quickly skimmed for relevant information, so small pieces of information can be retrieved quickly.
- Text can describe even objects that can't be shown in an image.

- Text can describe abstract concepts and actions (example: "let the tissue dry").
- Text can describe complex causal relationships (example: "if ... or ... and not ... then").
- Texts are easy and inexpensive to create.
- Texts are easy and inexpensive to update.
- Texts are easy and inexpensive to translate. (However, the fact that texts need translation at all is a disadvantage.)
- Texts can easily adapt to the size of the display on which they are shown.

Cons:

- With text, it's difficult to describe spatial properties and relationships clearly.
- With text, it's difficult to describe how an object exactly looks so that users can reliably identify the object.
- Texts are difficult to remember. Most people recall better what they have seen rather than what they have read.

Pros and cons of images

Pros:

- Images can reliably show even complex forms and spatial relationships.
- Images make it easy to identify particular objects.
- Images can avoid having to give an object a name. (Example: An image *shows* a toolbar icon instead of naming it.) Thus, users don't need to learn and remember lots of technical terms. Also, this reduces the risk of misinterpretation due to terminology issues.
- Images make it possible to read a document even when not sitting in front of the product.
- Images can attract attention.
- Images can motivate even those users to look into the documentation who typically don't read text.
- Images can be understood even by users who have only poor reading abilities.
- Images can be understood even by users who don't understand the document's language.
- Images provide relief on pages full of text.
- Images can work as a visual landmark for users who skim the documentation for relevant information. (Users typically look at the images before they read the words because it's more natural and thus faster.)

Which medium to use?

- Images are often perceived to be less tiring to consume. (With text, users need to construct some mental images themselves, which isn't necessary with given images.)
- Images are typically remembered better than text. (Example: Faces are remembered better than names.)
- Images don't need any translation, provided there's no text in the images. The more languages you have, the more important this advantage is. However, in case there *is* text in an image, translation is technically more difficult than with standard document text.
- Images prompt the coordination between the documentation and the product. When an image shows the result of an action, it prompts users to verify the success of their action, which else many users forget to do ("nose-in-the-book-syndrome"). When the image shows a step, it provides a clearly visible reentry point into the document.
- Other than videos, images also work on paper.

Cons:

- Images can only convey content that can be shown or visualized.
- In most cases, images take longer to create than text and thus are more expensive.
- Images can be difficult to view on devices with a small display.
- Screenshots may be mistaken for the real user interface so that users accidentally click them and nothing happens.
- In printed documents, when an image doesn't fit on the rest of a page, it creates a large gap (or the image moves to the wrong place in the document, which is even worse and should be avoided).

Pros and cons of videos

Pros:

- Videos can show objects and spatial properties very realistically.
- Videos are good at showing movements. A video can demonstrate a movement that the user needs to make even if describing this movement with words is very difficult. In doing so, a video can often communicate important details, such as how exactly a particular movement needs to be made, which physical forces are to be expected, etc. ("view over the master's shoulder").
- If a video has audio, it can combine visual and auditory learning.
- Videos are remembered particularly well because this is the way in which we naturally perceive the world.
- Watching a video often seems to be easier and more entertaining than having to read lots of text. Thus, many audiences are more likely to watch a video rather than to read an equivalent text.

Which medium to use?

- Many audiences are used to learning from videos from the Internet, so they explicitly look for videos.
- Videos sometimes can be used for both user support and as a marketing instrument.

Cons:

- Videos can be very slow at providing the desired information. Users may need to invest a comparatively large amount of time for watching an entire video or an entire scene even if they only need a tiny bit of information.
- Videos are bad at making specific information findable. Unlike text, the video cannot easily be skimmed for particular information. Information may not be discovered at all if users don't expect it to be in the video.
- The given information doesn't remain visible. It's difficult to repeat. For example, in a procedure users may have already forgotten how to perform the first steps of the procedure when the video has finished. To look something up again, the only way is to replay the scene up to the corresponding position (meanwhile forgetting the last steps ...).
- If a video uses audio, this may disturb other persons who are in the same room as the user.
- If the product is used (and thus the video watched) in a noisy environment, it may be impossible to understand the audio.
- If a video contains narration, and in particular if the video shows the presenter and his or her lips, translation is difficult.
- Videos don't work on paper.
- Users need a viewing device when using the documented product.
- Legal requirements may not regard video to be a sufficient form of documentation, especially in safety-critical contexts. So you might be able to use video only as a supplement to text, resulting in double work and cost.
- To be able to create a video, at least a prototype of the product must already be available. It's almost impossible to create the video only based on a specification, like it's largely possible with text-based documentation. (Exception: Videos that use animation.)
- Typically, videos are more time-consuming and more expensive to create than text.
- Typically, videos are more time-consuming and more expensive to update than text. So videos can get particularly costly for new products that are still under heavy development.

General recommendations

Studies have shown:

- With simple tasks, images in the documentation don't make users faster. Sometimes, too many images even slow users down in this case.

Which medium to use?

- With more complex tasks, images in the documentation *do* make users faster and avoid errors. Also, images help to better memorize the procedure.

Even though there are some exceptions, in general:

- New users watch videos and skip text.

    This is because new users mainly need task descriptions, which is the particular strength of videos.

- More advanced users skip videos and scan text (and may eventually even read some of the text).

    This is because advanced users often need to access only small pieces of reference information, which is the particular strength of text.

Thus, typically best use the following:

For ...	Best use ...
Tutorial	Video
Abstract concept	Text + still images
Simple task	Text
Complex task	Text + still images or video
Reference information	Text

- Don't use an image as an ancillary illustration of what you've already said. An image should *add* information but not just recite what's already in the text.
- Don't use an image to show what the product looks like. Good documentation does not describe what users can see by themselves. Good documentation explains what users can't see.
- Parts of existing documentation that are heavy with images—in particular with a series of images—are good candidates for video.
- The more likely it is that your audience has some good visual imagination, the more likely it is that you can well use still images rather than video. If your audience is engineers, you can typically presume that their visual imagination is above average.

With video and interactive content, think ahead about a replacement for print

Text and still images work in both online documentation and in printed documents.

Which medium to use?

If you use video in online documentation but want to provide the documentation also in printed form as a manual, you need to work out a concept of how to deal with the videos' content. The same applies to all sorts of interactive content.

You may either:

- A: Omit the video or interactive content in the printed documentation.

  This makes sense because you've already decided that in fact the best medium to present the content is just *not* text. However, there may be contractual or legal requirements that force you to provide everything in printed form as well.

- B: Replace the video or interactive content with text plus still images.

  The downside is that this means a lot of extra work: You need to write additional text, take pictures from the video, or even create new images. This significantly increases the costs of creating and translating the documentation and of keeping it up to date in the future.

Spend your budget wisely

When your resources in terms of time and money are limited so that you cannot create and add as many images and videos as you'd like to, maximize their total value. Use the most expensive media where they can unfold their maximum benefit. Add them to those topics of the documentation that:

- are most critical for the users' success
- are opened most frequently
- are viewed early while the users have not yet finally decided whether to hate or love your product

If you follow these principles, even if you only have a few images or videos, you have them where they make a big difference.

Related rules

Use each image for a clear purpose 49
Which type of video? 220
Interactive content 309

2.2 Common basics of visualization

The effectiveness and the usability of still images (drawings, photos) and of moving images (animation, videos) are largely determined by the same principles.

For technical documentation, the most important principle is the need for simplicity. You'll encounter this over and over again.

To start with, follow these tips:

Avoid visual noise [20]
Use color sparingly [25]
Use symbols sparingly [28]
Include text elements when necessary [33]
Tips on writing text within visuals [37]
Carefully consider whether to show persons [40]
Be aware of cultural differences [41]
Avoid legal pitfalls [44]

For more specific information, see:

Images in general [47]
Images of hardware [135]
Images of software [167]
Video design [219]
Video production [279]
Interactive content [309]

Common basics of visualization

2.2.1 Avoid visual noise

Don't overwhelm your audience. Remove everything from your images and videos that doesn't add any useful information. Leave out the visual "noise." Each object, line, and pixel that you can omit adds to simplicity and thus to clarity.

The simpler your image or video is, the better users can focus on its content.

Another advantage of simplified images and videos is that they can better be used for documenting multiple product versions.

Note:
Keeping images and videos simple doesn't mean that they shouldn't be visually appealing. Aesthetics is an important factor in motivating users, and it's an important factor for the perceived quality of your documentation.
However, a good image or video doesn't always have to be elaborately created and show every detail. The aesthetics of a good image in technical documentation basically lies in conveying the necessary information as simply and quickly as possible.

Examples

✘ No:

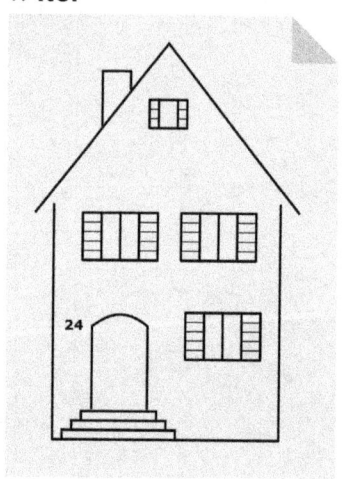

✔ Yes:

20

✖ **No:**

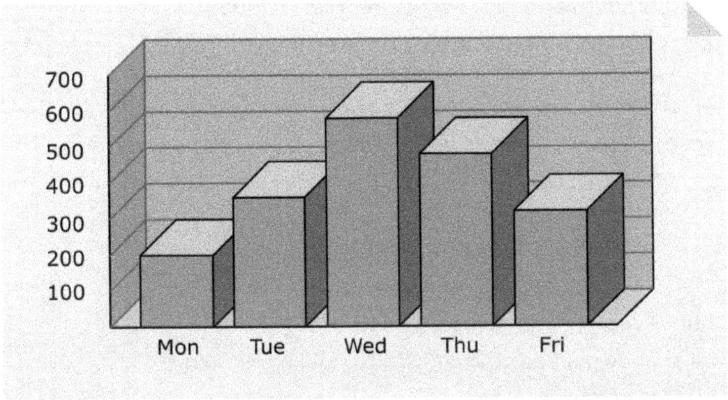

✔ **Yes:**

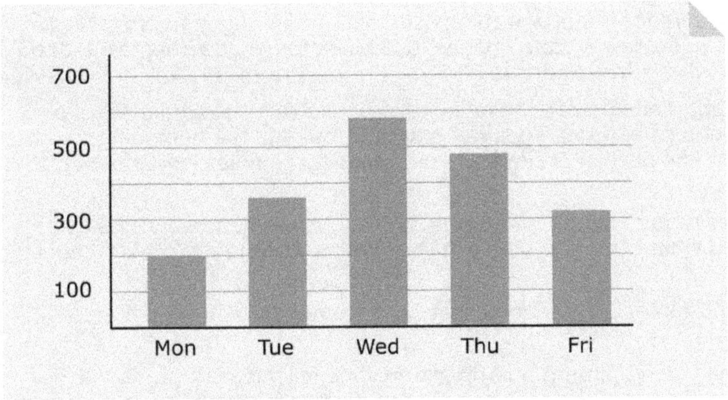

General ways of keeping an image simple

- Remove everything that isn't important for the purpose and message of the image.
- Arrange the remaining objects in a clear, structured way.
- Use enough white space to group objects that belong together.
- Simplify what can be simplified. Leave out unnecessary detail. (Note that what is "unnecessary" depends on both the audience and the purpose of the image.)
- Use as few lines and pixels as possible.

Common basics of visualization

- Remove repetitive information.
- Clearly distinguish the visual weight to separate primary information from secondary information (example: use different line widths).
- Keep all sorts of codings simple, such as color codings or codings by shape or line style.
- Increase detail only where it's needed to clearly identify critical items.
- If it makes sense, divide one complex image into two or more simple ones.

Particular ways of keeping a drawing simple

- Be as abstract as you can depending on your audience.
- Use simple shapes rather than complex ones.
- Don't use drop shadows unless this adds any significant value.
- Don't use 3D if 2D is enough. However, note that often 3D images are easier to understand for some audiences.
- Use only a few colors.
- Use bolder colors for important things and use gray or light colors for less important things. In grayscale images, use black for what's important, and use gray for what's less important.
- If it's sufficient, only *indicate* a line instead of showing the whole line. For example, it can sometimes be enough to only indicate the beginning and the end of a line. The human brain then automatically complements the rest of the line.
- Always consistently use the same styles for the same purpose. Once users have learned what each style means, this makes reading the other images much easier.
- As good as possible, stick to common conventions that your audience already knows.
- Don't use the same element and the same style for different purposes. For example, if you use arrows for indicating movements, don't use arrows also for pointing out objects in your image. If you can't avoid using the same elements for different purposes, at least give them a clearly distinct look.
- When importing CAD drawings from development, remove all things that are important only for internal use, such as approval signatures, headers, footers, etc. Also, you can typically remove a lot of unnecessary detail here to reduce the complexity to the level necessary for the users of the product.

Particular ways of keeping a photo simple

- Consider using a grayscale image plus only one additional color for highlighting objects or areas of interest.
- Crop the image so that only the relevant section is shown.

Note:
However, make sure to provide enough orientation. Don't cut off elements that users may need to see for being able to locate and identify the shown object.

- Erase objects that don't need to be visible. These may be elements that were needed only for taking the photo (such as holding hands or a supporting device), or it can be elements that don't add any useful information.

  Note:
  However, don't erase any objects that might be necessary to help locate and identify the important objects.

- Work with a shallow depth of field so that only the area of interest in the image is in focus but the irrelevant background gets blurred.
- Decrease the opacity or contrast of all objects that aren't the main object in the image.

Particular ways of keeping a screenshot simple

- Only show the window of interest but not the full application. With web applications, don't show the browser but only the web application itself.
- Don't make the window that is to be captured larger than necessary.
- Within the window, only show the relevant area (plus some surrounding context for orientation if necessary).
- Use simple sample data.
- If the captured window contains a list of uniform data, consider adding only a few sample data records instead of filling the window with a long list.
- Consider using a simplified user interface image (SUI) — see *Use a real screenshot or an illustration?* 174).

Particular ways of keeping a video simple

- If possible, remove all objects that aren't relevant for the message of the video.
- Only show the area of interest (plus some surrounding context for orientation if necessary).
- If the video shows software, use simple sample data.
- Use only few, unobtrusive video effects.
- Be careful with adding text and other elements, such as boxes and arrows.

For more information, also see *Keep the video simple* 257).

Common basics of visualization

> **Related rules**
> *Use color sparingly* 25
> *Carefully consider whether to show persons* 40
> *When to add borders around images?* 62
> *When to include the mouse pointer?* 183

2.2.2 Use color sparingly

> Colors are a powerful means of attracting attention.
>
> However, like having too many objects in an image, having too many colors creates a lot of "visual noise."
>
> The more colors are used, the less is the impact of each.
>
> Thus, use colors sparingly. In technical documentation, use color to *communicate*. Don't use color to *decorate*.
>
> Use the most striking colors for the most important things.
>
> Color can be used:
> - to mimic reality
> - to draw attention to something (or to distract from something else)
> - to distinguish things from each other or to assign them to groups
> - (for aesthetics and beauty)
>
> When possible, feel free to make colors beautiful, but never do it at the expense of clear communication.

Using color coding

You can use colors to assign the objects within an image to particular groups. Most color codings need to be learned by the user, so it makes only sense to use them if they are needed frequently.

For some colors and applications, there are already existing conventions. If you can, use one of these rather than to invent your own.

Examples:
- *red* = danger, *green* = OK, *yellow* / *orange* = caution
- *green* = biological, *blue* = water
- *blue* = for boys, *pink* = for girls
- *red* = hot, *blue* = cold

When defining your own color coding:
- As good as possible, base your choice on existing conventions known by the audience.

Common basics of visualization

- If there's no common existing convention, try to base the color choice on the color of real objects (example: green = forest; brown = field; gray = rock).

Use harmonic colors

People tend to find certain color combinations more pleasant than others.

On the Internet, you can find various tools that help you find groups of colors that go well together. This eliminates your personal preferences and can greatly speed up the selection process.

Some of the tools use manually compiled color palettes; others base their recommendations on the analysis of the colors of existing web sites or on automatic algorithms. Look for "color tool" or "color scheme designer." Typically, you give these tools one color to start with (for example, the main color in your company logo). The tool then returns a selection of colors that blend well with the given one.

Avoid colored text

Colored text often results in problems. Compared to black text, the contrast to the background is lower and therefore the text is more difficult to read. Under poor lighting conditions, the colors of thin letters are often hardly recognizable—so is the text written with these letters.

The problems get even worse if colored text isn't printed on a white or black background but on top of some other color.

If you *do* use colored text:
- Use it sparingly.
- Use dark, saturated colors.
- Use a rather bold typeface.
- Only use it on white, on black, or on shades of gray.

Consider poor lighting conditions

When designing your images, don't forget that many technical documents aren't read in a well-lit office but when actually using your product. This may be in a dark corner behind a machine or on a mobile device in bright sunlight.

Therefore, use colors that are bold and bright enough but not too bright.

Consider black-and-white printing

With online or PDF documentation, users may decide to print the documentation on a black-and-white printer.

If your images use color:

- Make sure that all colors produce a shade of gray that's clearly visible. This applies in particular to all light colors. For example, some light blue or light yellow may sometimes just look white on a self-made printout.
- Make sure that all colors produce shades of gray that are clearly distinguishable from each other. For example, in color some shades of blue and green may look clearly different. However, when printed on a black-and-white printer they produce exactly the same shade of gray. In the worst case, you do not only lose the differentiation between both colors on the printout. If the two colors adjoin each other in the image, two different objects visually merge into one and are then no longer recognizable.

Consider color-blind people

A significant proportion of men (about 8%) and women (about 0.4%) cannot reliably distinguish red from green. For this reason:

- Use color only as a second coding in addition to another coding, such as the shape.
- Don't use red versus green—or any shades derived thereof—for any critical distinction. (Traffic lights use the location top / middle / bottom as an additional coding.)
- Make sure that the brightness value of the colors is different so that people who can't see the colors at least can distinguish them.

Tip:
First make the image work in black and white. Only then, add color as a second coding to improve the image's performance.

> **Related rules**
> *Avoid visual noise* 20
> *Color or black and white?* 107

2.2.3 Use symbols sparingly

Symbols or "icons" are nonverbal abbreviations.

In technical documentation, you can use symbols:

- for warnings
- for other functional text elements, such as for tips and examples
- for navigation
- within images as a replacement for text

Symbols are also often visible on the product itself, such as on switches or on a software user interface. In these cases, you often need to show the symbols also in your documentation.

Like abbreviations and acronyms, symbols can speed up and facilitate communication. But symbols are also prone to be misinterpreted. Unclear symbols are one of the most frequent causes why users may fail to understand a manual correctly.

Use symbols only:

- if you're sure that all users know the meaning of the symbol or if you can successfully teach users what the symbol means
- if the symbol is needed so often that for the users it's worth the effort of learning its meaning

When in doubt, spell it out.

Advantages of symbols

- Known symbols can be recognized much faster than text.
- Symbols can be more memorable than text because they address the visual channel of the human brain.
- Simplified, small symbols can save space and reduce the amount of clutter on a page.
- Well-designed symbols can add to the visual appeal of a document.
- Symbols don't need translation (there may be exceptions with culturally dependent symbols though).

Disadvantages of symbols

- It takes much more time to develop a good symbol than to simply use text.

- Unknown symbols must be learned. Else, they don't have any meaning to the user.

> **Important:** Most users don't read technical documentation from start to finish. If you have explanations of symbols in the front matter or in the appendix of your document, these explanations will likely go unnoticed. Especially this is the case in online documentation, where users only access single help pages without seeing the rest of the documentation. Symbols that require explanation must therefore be explained immediately where they are used—possibly even several times within the same document.

Finding a good symbol

If there's already a standard symbol that the audience knows or that's common in their industry, use this symbol and don't invent your own.

Tip:
When looking for ideas on how to visualize something, you can use the image search of a web search engine. The results give you a good idea as to which images other people associate with your keyword. It's very likely that what you find here is what you should use because it's what's most intuitive to others.

A good symbol doesn't need explanation:

- When we look at a symbol, our brain begins to compare the symbol with the shapes of the objects stored in our memory. If there's a match, this is identified as the meaning of the symbol. Thus, when designing a symbol, you need to make sure that your audience actually knows the object that the symbol shows.
- Other symbols don't show real objects but common shapes like an exclamation mark or a traffic sign. These symbols only work if the users have previously learned what the particular shape means. Thus, when designing a symbol, you need to make sure that your audience already knows the meaning of the used shape.

Some symbols are even a combination of both: A simplified known object *plus* some abstract form with a known meaning.

When possible, use simplified images of concrete objects rather than abstract forms. Concrete objects are typically recognized more quickly, learned easier, and recalled more reliably than abstract forms.

Try to make use of some direct association, such as physical resemblance (mimicry). For example, use the silhouette of the object. Or make use of some association that users have already learned in their lives before. Be aware that both may be very specific to the audience and their culture.

Common basics of visualization

Examples for finding analogies and associations

Object / action	Analogy / association
document	physical analogy: sheet of paper
delete	associate an object: trash can
restaurant	associate an object that's a part of the whole: fork + knife
keep dry	associate a negated object: umbrella
edit	associate a tool: pen
left align	associate a result: left-aligned lines that symbolize left-aligned text

Examples for making use of established (learned) symbols

Symbol	Learned meaning
slashed circle	forbidden
triangle	caution
octagon	stop
plus sign	add / positive / good
minus sign	remove / negative / bad
x-mark	no / not / bad
check mark	yes / OK / good

Be aware of cultural differences

If your product is sold globally (or might be sold globally in the future), keep in mind that:

- Some objects may look different in other parts of the world.
- Some objects or symbols may be completely unknown in other parts of the world.
- Some known symbols may have a different meaning in other parts of the world.
- Some symbols may be interpreted differently in other parts of the world.

- Some gestures may be offensive or obscene in other cultures (especially gestures with hands).

Design tips

Make your symbols small and simple.

- Limit the size to what can be seen with one glimpse without moving the eyes.
- Combine a degree of realism with significant abstraction.
- Include as few objects and elements into the symbols as possible.
- Omit as many details as you can. Only show the most characteristic elements of an object. Every saved line and pixel count.
- Emphasize the single most critical characteristic. If you're developing a series of symbols, emphasize what distinguishes the symbols from each other.
- If possible, use only a two-dimensional view (3D needs more shapes and lines).
- Use simple, filled shapes.
- Make lines bold.
- Use only one color.

Develop a consistent family of symbols. Make all the symbols that you use consistent:

- Use the same size and background.
- Use the same colors.
- Use the same basic shapes and line widths.

Consider combining graphics with text.

If anything might be unclear to the audience, but you can't improve the symbol itself, add some short text as a hint. If the text is short, the symbol still retains its core benefits but avoids any doubt about the symbol's meaning. The downside is that the text adds some more clutter and may need translation.

You can also use text to add more strength (example: "DANGER") or to make an important distinction or specification (example: "ON" / "OFF").

For the usage of texts in images, see also *Include text elements when necessary* ⌈33⌉.

About using cartoons and comic characters

Comic characters are only rarely used in technical documentation, but sometimes they are—especially in printed getting started guides and in tutorials. If

Common basics of visualization

you can be informal, cartoons can be effective in "humanizing" and personalizing a document, in particular, for children and for non-technical audiences.

Cartoons can attract attention to important messages, such as warnings, tips, and reminders. Also, a cartoon can be a great means of showing the consequences of undesirable behavior in a drastic, yet inoffensive way. Bluntly telling users not to do something foolish can trigger the feeling that they aren't taken seriously ("I would never do something that stupid"). When using a funny character, however, this character takes the role of the idiot rather than the user.

Tip:
When using a cartoon, avoid giving the characters any personal characteristics, such as gender or ethnic group. Users should *not* be able to identify with the character. Better, animate real-world-objects or use a stickman. The simpler the better. Consistently use the same character throughout the whole documentation (where you use a comic character).

2.2.4 Include text elements when necessary

A good picture is not a picture puzzle. If needed, feel free to combine the visual coding with a verbal coding.

In technical documentation, many images and even some videos can be interpreted more easily if there's some short text included. From a didactic point of view, this is perfectly OK, but it has one major downside: If your documentation needs to be translated, images with text also need to be translated.

Essentially, the decision is a balancing of quality versus costs:

- If you need to minimize translation costs, don't use text in images.
- If you want to achieve the best possible quality, feel free to use text.

However, always keep the texts as short as possible. If you need much text, chances are that an image isn't the right medium.

A language-neutral (but often less user-friendly) alternative to adding text to an image can be the use of a numbered legend (see *Use text callouts or a legend for labeling?* 82).

Text-free instructions

To be completely language-independent, some manuals even go without any text at all. These manuals don't have any body text but consist of nothing but images. And also the images are free of any text.

For simple products, this can work well, such as for assembly instructions for furniture.

Typically, the images use a number of consistent symbols that represent particular tools, actions, requirements, or consequences. These symbols need to be crafted very carefully because there's no text that can compensate for any weaknesses here (for information on designing symbols, see *Use symbols sparingly* 28).

Use cases of texts in images

The following are some typical examples of where texts may be used in an image:

- **Name** the components of a device or of a user interface.
  Example: *Volume control*
- Add some brief information on the **purpose** of a component.
  Example: *Deletes all data from memory and restarts the device*

33

Common basics of visualization

- Add some brief information about the **behavior** of a component.
  Example: *Flashes red in case of an error*
- Add some brief information about how something **feels**.
  Example: *Press until you feel it click*
- Add some brief information about **prerequisites**.
  Example: *Can only be pressed while device is in maintenance mode*
- Add some brief information about **limitations**.
  Example: *Only available in countries of the European Union*
- Add some brief information about how **things work together**.
  Example: *Also activates power-saving mode*
- Add some brief information about special **properties and technical data**.
  Example: *Hose heat-resistant up to 250 °C*
- Provide some notes on how to **handle** an object correctly.
  Example: *Remove clamp before inserting*
- Give a **command**.
  Example: *Press here*
- Add corresponding **warnings**.
  Example: *CAUTION: Surface may be hot. Risk of skin burns. Do not touch.*

Combining text with a symbol

To emphasize particular information and to make it more findable, it can sometimes make sense to combine text with a symbol.

Examples:

🌡 **min 50 °C**

🎛 **10 kPa**

Common basics of visualization

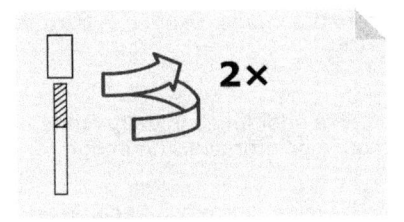

Procedures as an image

Sometimes, a whole procedure can be illustrated by just one image. Here it can be a very user-friendly solution to include all steps right into the image, pointing to the objects that need to be manipulated. Users then don't need to switch between some separate text and the image.

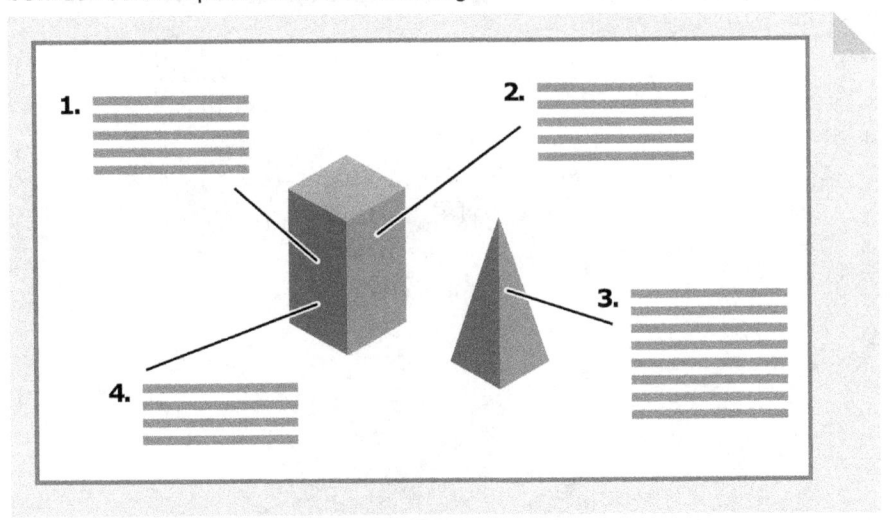

Note:
In steps 2 and 3, the end of the connecting line is placed near that part of the text that refers to the drawn object. In step 1, the line starts *below* the text because users should look at the drawn object only after reading the text. In step 4, the line starts *above* the text because users should first look at the drawn object to be able to better understand the text.

Alternative forms of adding text to images in online documentation

When creating online documentation, you have some advanced options for adding texts to an image. For example, you can only show the texts within a popup window when a user clicks a particular object in the image.

Common basics of visualization

- The advantage is that this removes clutter from the image. Users can better focus on the image itself.
- The disadvantages are:
  - It's not immediately obvious that there's some additional, initially hidden information. Users who don't move the mouse pointer over the image miss this information completely.
  - It requires some additional interaction by the user, which can become annoying and tiring.
  - It needs some extra work and special tools to create the effects.
  - It can be hard to use the same images also in a printed user manual or PDF.

Thus, this technique can sometimes be a solution for additional, optional information. However, for required, important information it's typically not the best choice.

Related rules

Use text callouts or a legend for labeling? 82

Tips on writing text within visuals 37

Tips on formatting image texts in general 95

Tips on formatting callouts and legends 85

You rarely need figure titles 101

You rarely need figure numbers 104

2.2.5 Tips on writing text within visuals

> When looking at an image or video, users aren't willing to read lots of text. Also, there isn't much space for long texts anyway.
>
> Write your texts in a way that makes it easy to skim them for the relevant information.
>
> Keep the texts simple, and get to the point quickly.

Minimize the number of words

Use as few texts as possible. Having lots of text in an image or video can be very distracting.

Keep the texts brief, both for clarity and for an uncluttered look of the image.

Move the important information to the beginning

When possible, start each text with the most characteristic and distinguishing keywords and information.

This makes it easy for users to quickly skim image texts for relevant content.

✘ **No:** *Here the operation can be canceled.*
✔ **Yes:** *Cancels the operation*

✘ **No:** *Sets the temperature*
Sets the pressure
Sets the exposure time
✔ **Yes:** *Temperature setting*
Pressure setting
Exposure time setting

✘ **No:** *Measurement – repeat*
Measurement – delete
Measurement – save
✔ **Yes:** *Repeats the measurement*
Deletes the measurement
Saves the measurement

Common basics of visualization

Keep the texts within the same image parallel

To facilitate skimming the texts, within the same image build all texts in the same way.

If possible, use the same syntax. Don't mix sentence fragments with complete sentences.

✘ **No:** *Temperature setting*
Adjustment for pressure
Sets the exposure time

✔ **Yes:** ***Temperature setting***
Pressure setting
Exposure time setting

Don't use unknown abbreviations

When there isn't much space available in an image, you may be tempted to use abbreviations and acronyms.

Only do so if you're sure that all users know the abbreviation or acronym. Abbreviations and acronyms are one of the major causes why many users fail with understanding technical documentation.

When in doubt, spell it out.

Don't be overly precise

When giving numbers, only add as many decimal places as you need to. However, be consistent when there are multiple numbers within the same image.

✘ **No:** *2.00 m*
3.00 m
1.00 m

✔ **Yes:** ***2 m***
3 m
1 m

✘ **No:** *2 m*
3.1 m
1.05 m

✔ **Yes:** ***2.00 m***
3.10 m
1.05 m

Use sentence-style capitalization

Capitalize only the first word of each text as well as proper nouns. Don't capitalize each word. (Note: This applies to the English language. Capitalization rules may differ in other languages.)

✘ No: *Remote Control*
Remote Control for Fan

✔ Yes: *Remote control*
Remote control for fan

Minimize punctuation

Add some end punctuation only if the text is a complete sentence or if it's a mixture of fragments and complete sentences.

✔ Yes: *Display*
Display to show current temperature
Display; shows current temperature
The Display shows the current temperature.

> **Related rules**
>
> *Include text elements when necessary* 33
> *Tips on formatting image texts in general* 95

2.2.6 Carefully consider whether to show persons

Showing persons is a very strong element in images and videos.

Show persons only if you have a good reason to do so—for example, to demonstrate how exactly an action needs to be performed.

Note that in some cultures, it may be offensive to show:

- persons in general
- particular groups of persons
- persons who aren't appropriately dressed in accordance with the customs of the culture

Tips and considerations

Our perceptual system treats images of persons with some particular priority. If a scene includes a person, we will inevitably notice and study the person first.

Thus, if the main subject in the image isn't the person, leave out the person or de-emphasize the person:

- If possible, don't place the person in the center of the image or video, where the attention is greatest.
- Reduce the size of the person by moving the person towards the background.
- In a drawing, reduce the level of detail of the person. For example, only suggest the face vaguely.
- Use light, unobtrusive colors for the person's clothing.

Related rules

Avoid visual noise [20]

2.2.7 Be aware of cultural differences

All sorts of visuals are particularly sensitive to cultural specifics.

When designing content for a global audience, take great care that your visualizations:

- can be understood as intended also in other regions of the world (and also by other ethnic groups within your own country)
- don't offend anybody

> **Important:** This is vital even if you don't have your documentation translated!

Possible issues with organization

Issue	Tips
Some cultures are used to reading from right to left. This does not only affect text but also: • the order in which users look at the objects within an image • the order in which users interpret a series of images	When in doubt, add arrows or numbers as a hint. These elements don't need to be large and dominating, but it can be helpful to have them.
Colors may have very different associations in different cultures.	Don't use color as the only coding. Consider using color only for highlighting but not for conveying a specific meaning.

Possible issues with objects

Issue	Tips
Objects that users are familiar with in your own country may be unknown elsewhere.	Avoid showing these objects if they aren't needed for the message of the image.

Common basics of visualization

Issue	Tips
Some objects may look different in other countries or regions (examples: power plugs, traffic signs, car dashboard depending on right-hand / left-hand traffic).	Try to omit the detail that accounts for the difference, or remove the object altogether. In a drawing, try to simplify the object, and use a more abstract shape so that it works for all variants.

Possible issues with persons

Issue	Tips
A gesture that seems totally clear may have a completely different, sometimes even offensive meaning in other cultures.	Don't use *any* gestures at all. Almost all gestures are obscene or rude somewhere in the world.
Showing particular groups of people may be offensive or even forbidden (examples: women, children, priests).	Omit or replace. In a drawing, simplify the shape so that it becomes neutral and shows just the outline of a person without any more detail.
Certain clothing (or the lack thereof) may be considered inappropriate or offensive.	Replace.
The images in your document show only persons of one particular ethnic group, color, or gender.	Mix to represent all user groups (but only these).

Possible issues with texts

These issues can not only occur in texts added on top of an image or video but also with data visible on the user interface of the shown product.

Issue	Tips
A mentioned law, national event, celebrity, movie, or other thing specific to your own country may be unknown somewhere else.	Don't use. These things will always fail somewhere.

Common basics of visualization

Issue	Tips
The name of a person may have an obscene or rude meaning in another language.	Replace.
The names of persons mentioned in sample data may represent only one specific country or ethnic group.	Mix names to include various languages and ethnic groups, and use both male and female names.
The decimal separator may differ. The form how dates are written may differ. The way in which prices are abbreviated may differ. The currency may differ.	If it's not critical for the message of the image, use your own locale. Normally, there's no need to hide your origin. If it's critical for the message of the image, you need to provide localized versions of the image. You may also be able to work with a locale that is generally known to the audience.
A month may belong to a different season.	Add a comment (example: "in Australia").

▶ **Related rules**
   Avoid legal pitfalls 44

2.2.8 Avoid legal pitfalls

With using images in published documents, there are quite a few legal pitfalls that need to be considered.

In order to avoid any trouble, pay close attention to the legal requirements:
- of your own country
- of all countries where you make your documents accessible

Things to pay special attention to

- If you're reproducing an image that was created by somebody else, make sure that you have the permission to do so. The same applies if you're reproducing only parts of another copyright holder's image, or if you're using another image within your own image. Note that many "free" images are in fact *not(!)* free to use for commercial purposes or may require proper attribution of their creator near the image or in your copyright notice.
- If a product, a building, a logo, a cartoon character, or any other copyrighted object is shown, make sure that you have permission to reproduce it.
- If an image or video shows a person, make sure that you have the written consent of this person to use the image or video commercially.
- If a video uses music, make sure that the music doesn't require any extra royalty payments based on the frequency of how often the video is viewed.
- Make sure that no confidential data is shown.
- Make sure that what's shown doesn't violate anybody's right to privacy.
- Make sure that your content may not be considered obscene in other countries.

Agreements with graphic artists and photographers

When working with graphic artists or photographers, it's important to know that payment for the creation of an image or photo does *not* automatically mean that you're allowed to use the works in your documentation or anywhere else! Unless the artist explicitly transfers particular usage rights to you, all rights remain with him or her.

Have the artist sign a written contract that clearly states who owns the right to use the image or video.

In particular, clarify:

- Do you receive exclusive rights of use or may the artist exploit the material in other ways as well (for example, also sell it to a competitor of yours)?
- Are you allowed to make changes to the image or video?
- Are you allowed to sublicense the use of the work (for example, if you sell your product and its documentation as an OEM product to other companies that sell it under their own name or integrate it into another product)?
- Are you allowed to create other works based on the material?
- Are your usage rights unrestricted or subject to certain conditions, such as crediting the author or limiting the use to certain countries, particular media, or to a maximum number of copies?

> **Related rules**
> *Be aware of cultural differences* 41

2.3 Images in general

No matter which type of product you're documenting and no matter whether an image illustrates a concept or shows hardware or software: Many principles apply to all images alike.

Essentially, it's all about making your images plain and simple. Users appreciate clear images. The facts are usually complicated enough.

Also, it's important that your images are designed and structured consistently throughout the entire documentation. This adds even another layer of simplicity because once users get used to the basic design they can fully concentrate on the content in all subsequent images.

With images, follow these basic rules:

Use each image for a clear purpose 49
Position each image precisely 51
Page layout of pages with images 55
Balance the visual weight of images with headings 60
When to add borders around images? 62
What image size? 64
Indicate the size and context of the subject 67
Guide the users' eyes 70
Emphasize what's important 77
Zoom in on details 79
Use text callouts or a legend for labeling? 82
Tips on formatting callouts and legends 85
Tips on formatting image texts in general 95
You rarely need figure titles 101
You rarely need figure numbers 104
Color or black and white? 107
Which resolution? 111
Which color mode? 116
Which file format? 118
Think ahead about editing an image 121
Think ahead about reusing an image 123
Think ahead about translating the texts in an image 125
Tips for developing graphics on your own 128
Tips for working with graphic artists and photographers 131

Images in general

Also note the general rules of visualization, which apply as well. See:
Common basics of visualization 19

For more specific information on particular image types, see:
Images of hardware 135
Images of software 167

2.3.1 Use each image for a clear purpose

It may sound trivial, yet it's the most frequently made mistake with images:
- Don't add an image to a document only because you think it "looks better" to have an image there. And even if it actually does look nice: If the image has no other purpose than being beautiful, it just stands in the way of the users.
- Don't add an image only because you feel that you need any particular number of images. If your document or parts of your document don't need many images, this is fine.
- Optimize each image for exactly one purpose.

Standard image purposes

An image can:
- **Help develop a mental model**.

    A mental model is the users' understanding of how a product works, how its controls are organized, and what the logical flow is when working with the product.

    The mental model helps users:
    - to construct strategies for action—which is important for being able to handle novel situations
    - to understand why an action produces a particular result—which is important for problem solving

    With software products, developing a mental model is often particularly important because users can't examine any physical components to build their own model.

    With software, you often need an illustration or a flow chart to explain the mental model. With hardware, it can often help to show some of the components that are hidden inside the product.
- **Locate and identify components and controls**.
- **Demonstrate how to perform an action**.
- **Help to determine the state of the system** (prerequisites before performing an action; results after performing an action).
- **Work as a landmark** within the document so that users can find content by just browsing the images.
- Only rarely in technical documentation: **Decorate, motivate, set a tone or mood, arouse curiosity, portray aesthetics of the product**, or just **provide visual relief** by breaking up long blocks of text. (If your document

Images in general

needs some visual relief, better use subheadings, lists, tables, and other functional text elements for this purpose. These elements also improve the structure of the document.)

Tip:
In your documentation plan or documentation style guide, define a separate image class with specific usage rules for each image purpose.

Don't make yourself a slave of frequency

Don't set up or stick to any rule that asks you to add a specific number of images, such as "at least one image per chapter" or "not more than one image every two pages."

This does not make any sense. Some manuals need many images, other manuals don't need any images at all. Also, not every part of a manual needs the same density of images.

Add an image only where the image adds value.

- An image that doesn't add any value is nothing but wasted space and extra clutter that users need to filter out to get to real information.
- With hardware products in particular, images can be quite time-consuming to create and thus expensive.
- With software products in particular, it can be costly to keep images up to date and to translate them (see also *How many screenshots to show?* 169).

2.3.2 Position each image precisely

Don't let the images positions be determined by the layout.

Other than, for example, in a sales brochure, in technical documentation images aren't artwork but a core element of the information. This information needs to be included at the exact right position.

- Add an image exactly where it's needed. This makes clear where the image belongs, and it enables the users to read the document continuously without needing to search for corresponding images.
- Don't put an image away to another page or to another place only because there is more space.

If there's some white space before or after an image, so be it

When you aren't flexible with the exact position of an image (you shouldn't be flexible), in printed documentation it sometimes happens that an image is too large to fit onto the rest of a page. So the image needs to start on a new page, leaving some blank space at the bottom of the previous page.

This is OK. Do *not* move the image anywhere else for this reason.

- In the text, it would make it difficult for the user to find the right image.
- Most technical documentation needs to be updated frequently and also needs to be translated into other languages. Some documents are also published in different variants for different audiences or product versions. Because the lengths of the texts will differ in each document, so would the tweaked positions of the images. You would need to invest lots of time to tweak this manually each time for each document. (You may get some automatic support from your authoring system here, but this can move the images even further away from the places where they are actually needed.)

Images in general

✖ No:

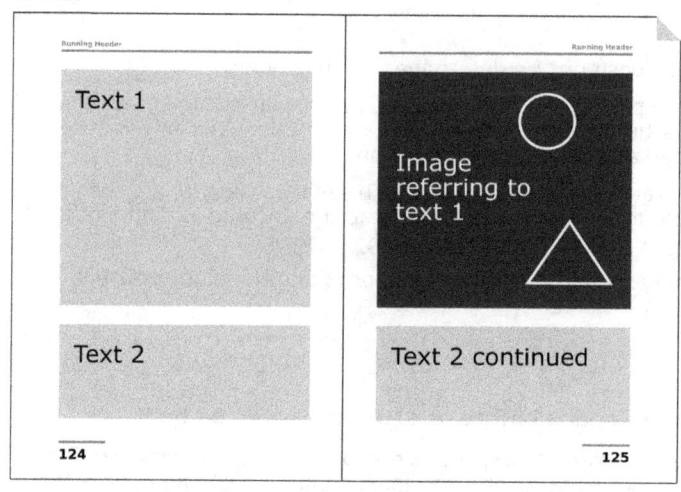

✔ Yes:

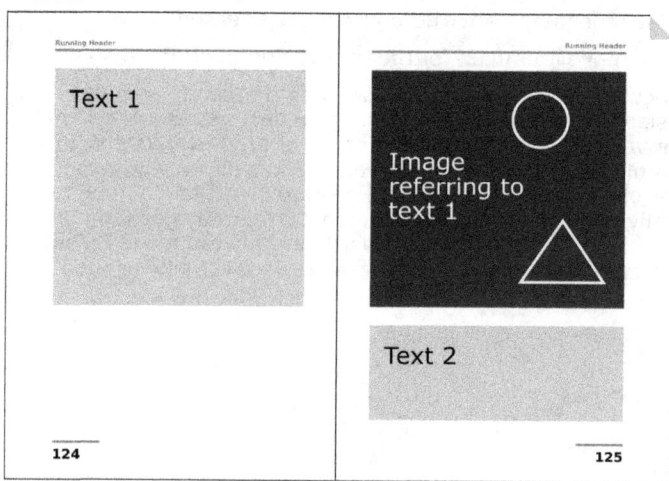

Best image position in descriptive text

- In descriptive text, include the image after you've first mentioned the fact that the image illustrates. Users can then interpret the image correctly. If you put the image before the description, users wouldn't know what the image is about when they first see it.

52

- If some longer explanation is needed, put the image between an introduction and the explanation. This is particularly important if the explanation can only be understood with the help of the image.

Best image position in procedures

- If the image shows the initial state of the product, place the image after you've described this initial state or after you've mentioned the prerequisites of the procedure.
- If the image shows an action, place the image after the text that describes the action.
- If the image shows the result of a single step, place the image immediately after the description of this single step.
- If the image shows the result of the procedure as a whole (that is the final state of the product), place the image after the whole procedure.

Feel free to repeat an image

If the same image helps in various places, don't hesitate to add it multiple times. This is perfectly OK. The improved user-friendliness is well worth the extra space that the image costs you in a printed user manual. In online documentation, redundancy doesn't matter anyway.

Don't presume that users have already seen the image before. Users typically don't read documentation from start to finish. So each place where the same image is included may be the first one for the user to come to. Having the image right in place then is much faster and provides a much better reading experience than either not having any image at all or than having to follow a cross-reference to the image.

Tip:
If your authoring system supports it, don't embed a copy for each instance of the image but include the image by linking to a unique image file. Then, should you later need to edit the image, all places where the image is used are updated automatically.

Exception: Images that primarily act as a reference

If images aren't necessary for understanding the text but are primarily used as a reference (such as charts for looking up values), it's often best to put them into a separate section of a manual where they can be quickly accessed in a targeted manner.

Another place where it can make sense to have a particular reference image is the beginning of a manual. This is often the case with overview images that outline the user interface of a product or its components. Because the image is at the very front of the document, users can access it easily and quickly at any

Images in general

time. There, more space is available and the image is always visible after flipping it out. In addition, the double-thick page can be found quickly just by feeling it.

▶ **Related rules**
Page layout of pages with images 55

2.3.3 Page layout of pages with images

In most cases, it's best to use a layout with *alternating* texts and images. Depending on the size and level of detail of the images as well as on the paper size, there may be one or several columns.

For procedures that need many images, sometimes a good layout is a two-column layout with *opposing* texts and images.

In online documentation, you almost always have a single-column alternating layout. In case you have a two-column layout with opposing texts and images in online documentation, it needs to automatically transform to a single-column layout on devices with a small display (responsive design).

However, regardless of the layout, always remember not to include more images in your document than necessary (see *Use each image for a clear purpose* 49).

Alternating texts and images

This is the basic layout, which is most often used. In particular, it's most often used for software documentation because screenshots of software typically have many details, which requires the full paper width or screen width.

Tip:
In online documentation, using clickable scaled-down image thumbnails can be an excellent option to save space and to make even a long topic with many images look very clear (see also *What image size?* 64).

Images in general

As a variant, you can have the text flow around an image (provided that your authoring tool supports this feature).

The advantage of text flowing around images is that it can save space and looks more dynamic, especially with small images.

The disadvantage of text flowing around images is that the images are visually not as precisely positioned as with alternating text and images. When there are multiple images each followed only by some short text, the layout may become quite disrupted.

Therefore, this technique is rarely used in technical documentation.

Two-column layout with opposing texts and images

When providing step-by-step instructions that need many images, a good solution can be to have two columns: One column with images, and a parallel text column with the descriptions of the steps.

There can be just one step next to an image, or there can be multiple steps if they don't need an extra image.

Images in general

Typically, this approach makes the most sense in a printed manual where you have a given page size. If you want to use the approach also in online documentation, you need to make sure that either:

- all users use a specific screen size (which can be the case if you supply the viewing hardware—for example, if the documentation is viewed on a monitor attached to a machine)
- the design is responsive so that on smaller screens or if the window has been resized to a small width, the images move into the text column and then alternate with the texts

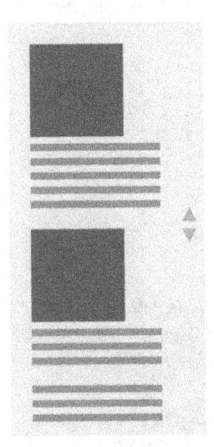

Option: Text fused with the images (free layout)

If your document has only little text but a very high proportion of images, an option can be to position the images freely on the pages. For example, you can find this often with printed setup and assembly instructions.

This approach, however, typically doesn't work with online documentation, or it requires some very sophisticated responsive design.

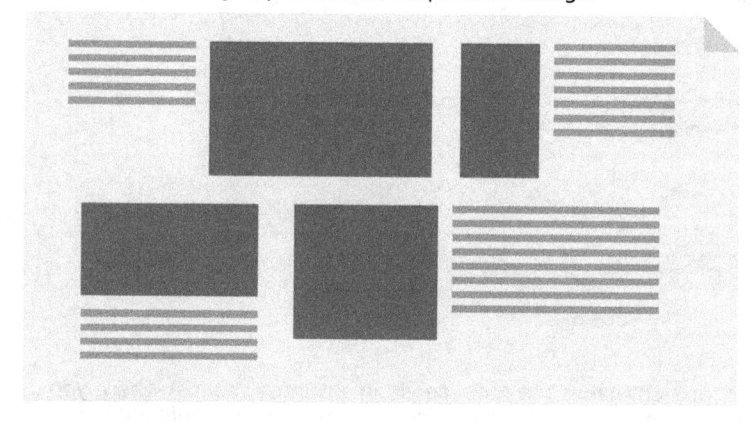

Don't make yourself a slave of pagination

In technical documentation:

- Don't crop an image just to get a nicer page break.
- Don't resize an image just to get a nicer page break.
- Don't even think of moving an image to a different position just to get a nicer page break.

In technical documentation, having a pixel-perfect layout with no white space at the bottom of a page is clearly less important than having perfect content. However, if you move an image to another position than where it best fits, or if you manipulate an image only for the sake of beauty, your content is no longer perfect.

Also, manually tweaking page breaks can become very time-consuming and expensive in the long run:

- Typically, technical documentation needs to be updated frequently. When you add or delete content, subsequent page breaks also change. Manually modified page breaks not only become ineffective then but may even produce very undesired results.
- When the documentation is translated, words in other languages may be significantly longer or shorter than the original text, which also shifts your page breaks. You would need to tweak everything again in each language.

Better create a good template that automates page breaks as much as possible, and then live with the outcome. There are certainly things in your documentation on which you can spend your time and budget more wisely than on manually tweaking pages breaks with each version and language of the documentation over and over again.

Images in general

Related rules

Position each image precisely 51
Balance the visual weight of images with headings 60
When to add borders around images? 62
What image size? 64
You rarely need figure titles 101
You rarely need figure numbers 104

Images in general

2.3.4 Balance the visual weight of images with headings

> Images attract much attention. Make sure that the visual weight of the images doesn't dominate the visual weight of the document's headings.
>
> Users must still be able to recognize the overall structure of the document easily.

Example

The following example shows a page that obscures the structure of the document by overly dominating images as opposed to a page that makes understanding the document structure easy.

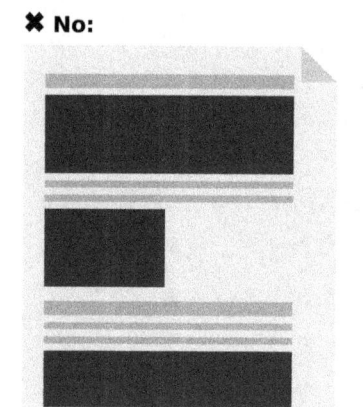

✘ No: ✔ Yes:

Ways of decreasing the relative visual weight of images

To make sure that images don't visually dominate a document:
- Avoid using bold colors in the images.
- Avoid using very bold, black lines in the images. Rather, just use shades of gray for bold lines.
- Avoid using large dark and large color-intensive areas.
- Don't use more images than necessary.

- Don't make the images larger than necessary.

Ways of increasing the visual weight of headings

To increase the visual weight of your headings so that they dominate a page:
- Use a large font size.
- Use a bold font style.
- Use color instead of black or gray.
- Add more space above the headings.

> **Related rules**
> *Page layout of pages with images* 58
> *When to add borders around images?* 62
> *What image size?* 64

2.3.5 When to add borders around images?

Most images don't need borders.

- More often than not, image borders only add some extra lines to the image without adding any extra value. This makes the document page that embeds the image more cluttered and thus often more difficult to read.
- However, if you're going to embed multiple images on the same page, image borders may help to separate the images from each other clearly.
- If an image contains any text, a border can also help to clearly separate the image from the body text of the page. So you need less white space around the image.

Tip:
An elegant alternative to using borders can be to add some light solid background color to the images.

Screenshots typically don't need any borders because they usually already have some background color or window frames visible in the image, which clearly define where the image ends.

Tips for designing image borders

In case you *do* decide to add image borders:

- Use a narrow line width. Typically, the width should be about the same as the width of the letter "I" in the document's body text.
- For the color of the lines, consider using a shade of gray instead of black. This looks less obtrusive, yet it's strong enough to guide the human eye.

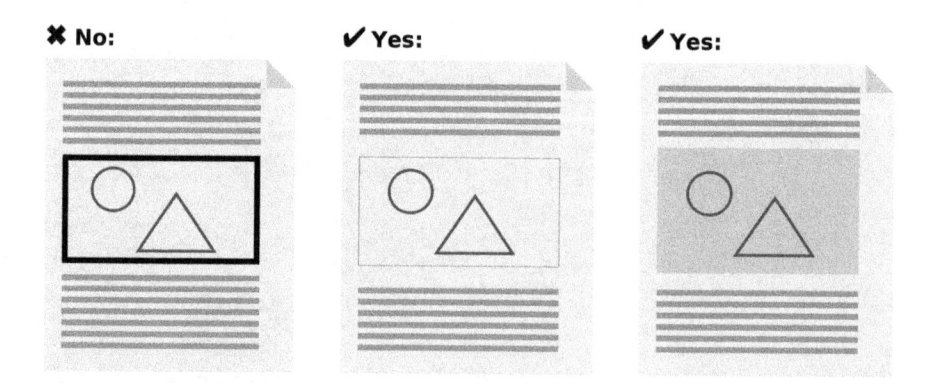

✖ No: ✔ Yes: ✔ Yes:

Using incomplete borders

To make a page visually more appealing and to avoid as much clutter as possible on an image and on the page that embeds the image, you sometimes can:

- use only one horizontal borderline or one vertical borderline in order to separate an image from another image
- interrupt a borderline and let a part of the image protrude from the frame
- not draw the complete borderlines but only indicate the corners and a small section of each line

✔ **Yes:**

Related rules
Avoid visual noise [20]
Page layout of pages with images [55]
Balance the visual weight of images with headings [60]

2.3.6 What image size?

> Simply put: Make images as small as possible but as large as necessary.
>
> In technical documentation, images aren't for the beauty of the document but for providing information. Smaller images have several advantages here compared to larger ones:
>
> - Reducing oversized images puts more information in the user's field of view. Users need to turn fewer pages or need to scroll less often on screen. This improves the usability of the documentation.
> - If an image is small, it can be viewed at a single glance without moving the eyes. This is less tiring and faster.
> - If there are many large images, they can visually dominate the document so that users don't see the structure of the document anymore.
>
> **ⓘ Important:** When scaling down an image that has been embedded into a document, along with the image you also scale down all texts and line widths in the image. Make sure that all texts remain sufficiently large to be easily readable and make sure that no lines become so thin that they disappear.

Tips for optimizing the image size

- An image typically is large enough if it can be read from the same distance as the rest of the document.
- When in doubt, print it out. When creating printed documentation, you just sometimes need to print out a page and take a look at it from the users' perspective to decide whether the image is clear and where you might further improve it. Don't rely on the preview on screen.
- If an image is no longer readable after resizing it to the required size, try to subdivide it. (In case you cannot subdivide the image logically, slightly overlay the coverage of the two halves to tell the users of the split and to help them identify where the one image ends and the other one begins.)
- Often, not all details of an object need to be visible in an image. For example, if it's the purpose of an image to identify an object, it may be sufficient if the shape and position of the object become evident. The image may not need to be so big that users can see irrelevant details.
- Often, not all objects need to be visible in an image. Many images can be significantly simplified and reduced in size by cropping out unimportant things. Note, however, that it must still be possible to identify the size and

the position of the shown objects (see *Indicate the size and context of the subject* 67).

Using thumbnails in online documentation

If it's not important to see all the details of an image at first sight, consider embedding the images as a clickable thumbnail.

When a topic is opened, the image thumbnails give an idea of what each image is about, but they don't take up much space. The structure of the whole page remains clearly visible. Users won't need to click and open many of the thumbnails at all—either because they don't need the information, or because they can already see enough from the thumbnail.

As with full-scale images: Make the thumbnail images as small as possible but as large as necessary so that users at least get an idea of what the image shows. This prevents them from clicking on images that finally don't provide the expected information.

Where consistency matters—and where it does not

Usually, there's no good reason why all images should have the same size. A detailed image may need the whole width of the type area to be clear, whereas an image with less detail would be way too big then.

However, there are cases in which identical sizes of the shown objects are important:

- if several images belong to a common sequence and show the same object (example: an object is manipulated in multiple steps)
- if several images show objects whose sizes are important or need to be compared

Tip:
To achieve some consistency in the document even when having different image sizes, you can use a selection of standardized image sizes, such as 50%, 75%, and 100% of the width of the type area.

Images in general

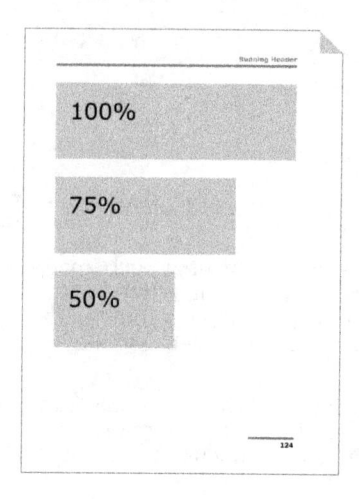

Related rules

Page layout of pages with images 55

Balance the visual weight of images with headings 60

2.3.7 Indicate the size and context of the subject

Important best practices about images are to show only what's important in an image and to keep images as small as possible.

However, there's one crucial extension to these rules: When showing a component of a device or a section of a user interface that users do not yet know, include enough marginal information to make clear:

- How big is the object?
- Where is the object located?

For example, when describing a procedure that involves the removal of a particular protection cap, don't only show the isolated cap. Even though the characteristic shape of the cap already contains helpful information, finding the cap on a complex machine may be hard. However, if the image also visualizes how big the cap approximately is and which known or easily recognizable objects are nearby, this makes finding the cap much easier.

The same applies, for example, when users need to find a particular button on an operator panel or within a software user interface.

Methods for indicating the size

For example, you can:

- include a person or a hand in the image
- include an object that has a well-known size, such as a wrench or a screwdriver
- include a tape measure or a folding rule
- include dimension lines

Use dimension lines only if no other method works, if you need to be very precise, or if your audience is very technical. Dimension lines can add a lot of clutter, and some audiences may have trouble reading them correctly.

Methods for indicating the context

For example, you can:

- include some close objects that users already know
- include some close objects that can easily be found and identified

Images in general

(in screenshots these are, in particular: window borders next to the area shown, or visually dominant and characteristic elements next to the area shown – see also *What to show in a screenshot?* 177)

- make the entire image smaller and add a magnifying effect (see *Zoom in on details* 79)

✘ **No:**

✔ **Yes:**

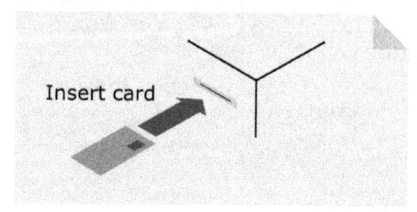

✔ **Yes:**

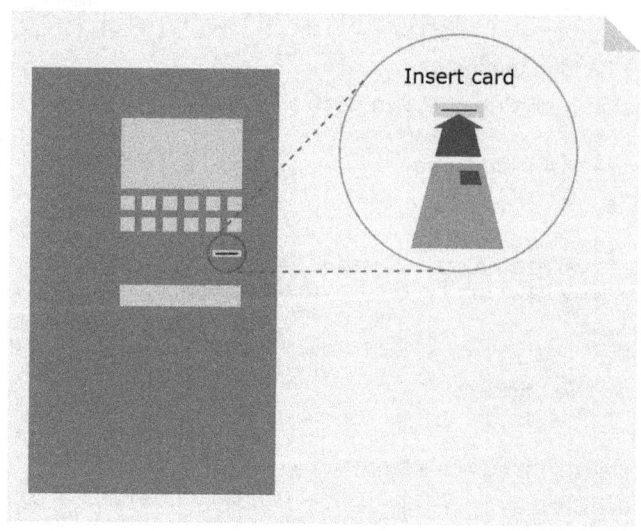

Tip:
If only a minority of users will need the contextual information in the image, this information is clearly less important than the main object in the image. Thus, consider de-emphasizing the contextual information by reducing its color saturation or by using a smaller line width (see also *Emphasize what's important* 77).

Tips for optimization

The disadvantage of showing lots of marginal information is that it makes your images bigger and more crowded. The following techniques can help you to minimize the impact of this:

- Often, you don't need to show it all. It may be well enough to show only some part of the neighboring objects.
- Often, it's enough to indicate the neighboring objects with just a few lines.
- To some degree, you may blur the background or display the background in lighter colors. So it's clearly subordinate to the object shown, but it still provides sufficient information about the context and size.

2.3.8 Guide the users' eyes

Control the users' attention. Add visual cues to guide the users purposefully through your image. This helps them to:
- save time
- better understand the image

To guide the users' eyes, you can make use of the following mechanisms and elements:
- In cultures that are used to reading from left to right, the users typically first look at the upper left corner of an image.
- If there's a visually very strong element in the image, the users will start there.
- If there are multiple visually strong elements in the image, the users will look at these elements one after the other.
- Things that can make an element a visually strong eye-catcher, are: darkness, color, size, and particular shapes that we are programmed to look at, such as human bodies and faces.
- Apply the so-called "Gestalt Principles" (see sections below):
    - Use the *principle of proximity* to visualize what belongs together.
    - Use the *principle of similarity* to point out what works equally.
- Use lines to guide the users' attention to a particular object or from one object to another.

Users have also learned that numbers and arrows typically indicate a sequence. So you may also use numbers and arrows for guiding the users along a particular path. However, note that arrows and numbers don't necessarily immediately catch the eye. This only happens if they are bold and large enough and stand out from the other elements in the image.

Tip:
Consider making the first number or the first arrow bolder than the others. This immediately indicates where to start. All other numbers and lines can be designed less obtrusive.

Using the principle of proximity for grouping elements

The closer the elements in an image are clustered together, the stronger is our impression that these elements belong together.

You can:

- reduce white space to group objects that are related
- add white space to separate objects that *don't* belong together
- if space is limited: use lines to separate objects that don't belong together but can't be moved further apart in the image
- use frames to group objects that belong together

However, note that each line or frame adds more clutter to the page. If possible, better use white space around a group, or add a light common background area to a group instead of using a frame (place the group on a filled rectangle or on another shape).

Tip:
Lines often don't need to be complete in order to guide the eye as intended.

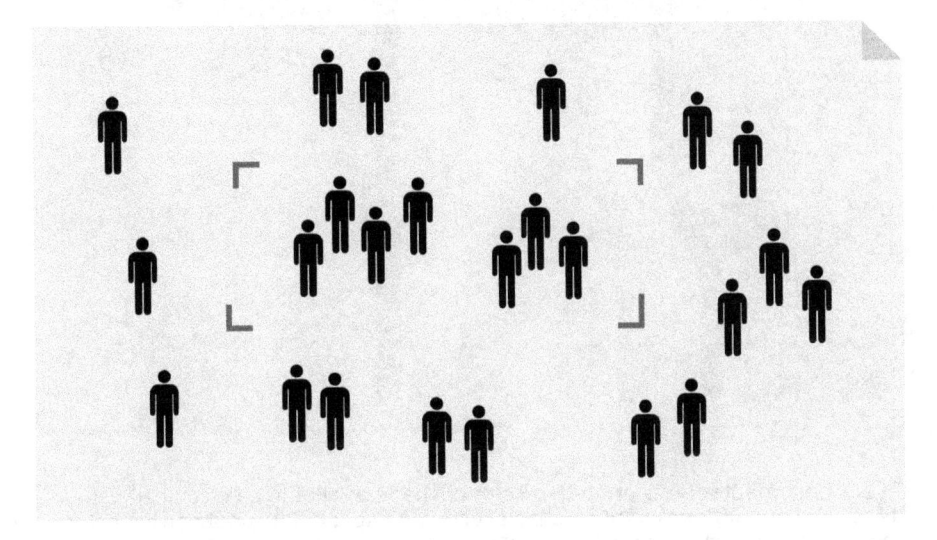

Using the principle of similarity for grouping elements

The more properties some elements in an image have in common, the stronger their visual coherence is.

Thus, if you want to visualize that some elements belong together or need to be handled in combination, give them identical:

- colors
- textures
- shapes
- symbols

Images in general

- fonts
- size

Application example:

For example, in an image you may use 2 different shades of gray for categorizing the components of a system:

- light gray for the entire module
- dark gray for emphasizing the particular components that are involved with the current step of a procedure

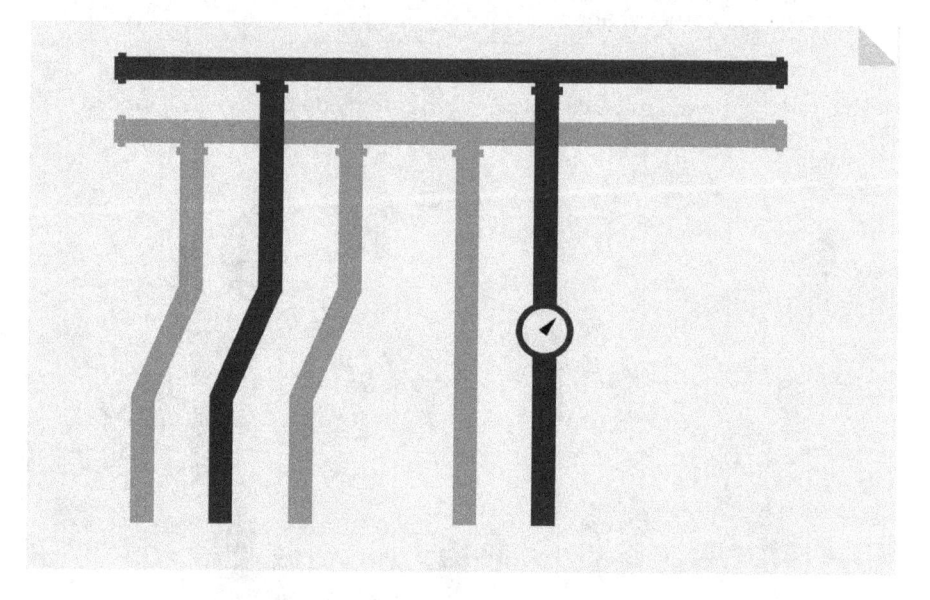

If your documentation is printed in color or made available electronically, instead of using dark gray for emphasis, you can use a single unobtrusive color. This can produce a very professional and stylish, yet uncluttered look.

- Use the same color consistently in all images.
- Chose a color that blends well with the overall design of your documentation, with the design of the product, and with your corporate style.
- Chose a color that when printed on a black-and-white printer produces a darker shade of gray than is the shade of gray of the other colors.

Images in general

Additional methods for grouping elements

In addition to using space and common characteristics for grouping elements, you may also:

- connect the objects that belong together with a line
- arrange the objects in simple geometrical ways, such as in rows and columns, or in a circle

Making use of the visual weight

We tend to look at those things first that have the biggest visual weight.

An object's visual weight increases:

- the closer the object is positioned to the upper left corner or to the center of the image
- the larger the object is
- the more colorful and the darker the object is

Put the most visual weight on:

- the objects that users should notice first
- things that you want to emphasize

Decrease the visual weight gradually while guiding the users' eyes from object to object (for techniques, see *Emphasize what's important* 77).

In complex images, you can also use the visual weight to layer the image according to importance: Primary information gets more visual weight than supplemental information.

To test the visual hierarchy of the objects within an image, you can make the "squint test":

1. Look at the image at its typical size like a user would do.
2. Close your eyes.
3. Slowly reopen your eyes. The first thing that you notice or recognize is the most conspicuous thing in the image (has the most visual weight).
4. Continue opening your eyes and observe the order in which you notice the other objects.

Images in general

Designing a visual path

Best, place the items in the image in a way so that they give the user a continuous visual path from the upper left toward the lower right as much as possible.

- Try to place those items close together that are processed one after the other, or that users need to compare.
- Try to place the most important objects near the center of the image.
- Visualize any sequence with connectors, arrows, or numbers.

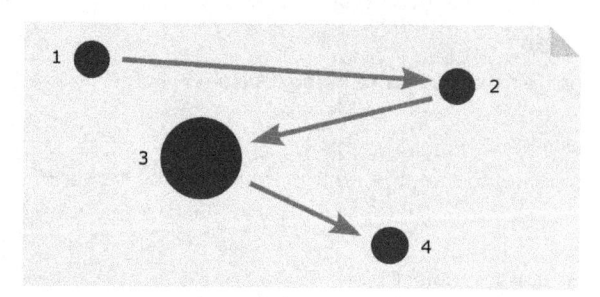

Tip:
Arrows don't always need to look like clumsy typical arrows. Sometimes, you can just use some subtle arrowhead-like shades in the background to direct the users' eyes.

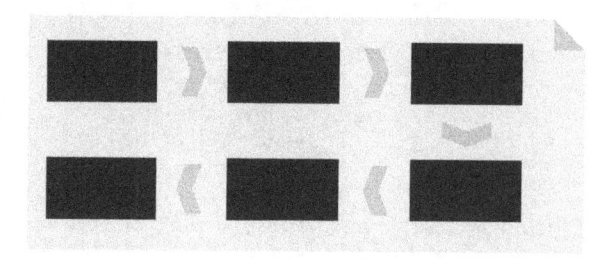

Relative positions and their associated meanings

By learned conventions, different positions in an image typically have certain associated meanings:

Images in general

Top
beginning
input
lightweight
rare

Background
distant
unimportant

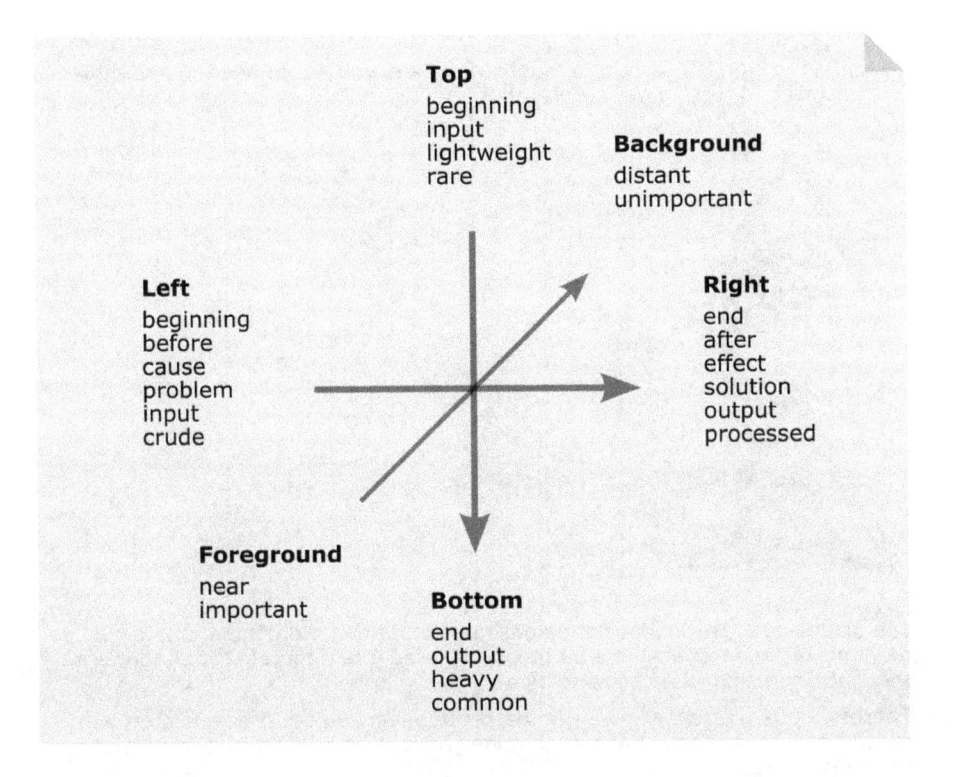

Left
beginning
before
cause
problem
input
crude

Right
end
after
effect
solution
output
processed

Foreground
near
important

Bottom
end
output
heavy
common

> ⓘ **Important:** Be aware of cultural differences. The users' accustomed reading direction greatly affects the direction in which they read an image, as well as the direction in which they interpret a series of images. In countries where the dominating reading direction is from right to left, some users might even be used to *both* reading directions. So determining the actual direction isn't always easy. When in doubt, add a second coding, such as arrows or numbers, to make the direction clear.

For example, the following traffic sign means that the road will go *up*. However, unless users know the convention that the sign is to be interpreted from left to right, they may just as well think it means that the road will go *down*.

75

Images in general

To make the sign fail-safe for a global audience, you would need to add a hint to the correct direction of interpretation:

The downside of these alternative designs is that the additional graphic element makes the traffic sign a bit more cluttered. So it may take a bit longer to recognize the sign when passing by quickly.

Therefore, it's ultimately always a matter of balancing the pros and cons with respect to your particular audience and with respect to the situation when the image is viewed.

Related rules

Emphasize what's important 77

Zoom in on details 79

2.3.9 Emphasize what's important

Show the user what to pay attention to in an image.

In general, you have the following options:
- You can emphasize the important items.
- You can de-emphasize the unimportant items.
- You can remove the unimportant items (thus showing only the important ones).

> **ⓘ Important:** When removing items, don't remove any items that users might need for orientation (see also *Indicate the size and context of the subject* 67).

Methods for emphasizing

To emphasize an object within an image, you can:
- Place the object near the center of the image or near the upper left corner.
- Use color on the object.
  - On a black-and-white image, use a soft color.
  - On a color image, use a bright color that's significantly different from the main colors in the image.
  - If your document is printed in black and white, instead of using a color, you can use a shade of gray that's darker than the surrounding objects (see also *Guide the users' eyes* 70).
- Add a frame or a circle around the object.

  For the lines, best use a solid color.
- Add some colored background behind the object.
  - Best use a subtle color because stronger colors are extremely strong and dominant over a large area.
  - Make sure that the background is also visible when the document is printed on a black-and-white printer.
- Add an arrow that points to the object.

  For the arrow, best use a solid color but no contour lines. This adds the least clutter to the image.
- Enlarge the object or add a zoom effect, such as a zoom lens or a zoom box.

Images in general

For examples, see *Zoom in on details* 79.

- Add more detail to the object than to other objects.
- Add more space around the object than around other objects.
- Increase the line width of the contour lines of the object.
- Add a symbol next to the object, such as a warning sign.
- Add some text next to the object, such as "IMPORTANT" or "CAUTION."
- In online documentation: Use some animation.

However, be very careful with animation effects. They are extremely strong. In fact they can be so strong that they become annoying.

> **Important:** Don't emphasize too many objects. Else, the effect is lost and the image just becomes confusing with so many highlights on it.

Methods for de-emphasizing

To de-emphasize unimportant objects in an image, you can:

- Blur the objects.
- Decrease the contrast of the objects. (If your image editor doesn't support this, you can overlay it with a semitransparent, white, filled rectangle.)
- Decrease the line width.
- Decrease the size.
- Use a light shade of gray instead of color.
- Use a light shade of gray instead of black.
- Use dashed lines instead of solid lines.
- Crop an object so that only a part of the object is visible.
- Add some text next to the object, such as "(optional)" or "Only needed in case …."
- If possible: Add some kind of brackets around the object.

> **Related rules**
> *Guide the users' eyes* 70
> *Zoom in on details* 79

2.3.10 Zoom in on details

Adding a zoom effect to an image can often do a very good job for several reasons:

- It points out which is the relevant section of the image.
- It makes the relevant details of the image large enough so that even visually impaired users can see them.
- Even though the image provides enough detail it can be relatively small.
- Even though the image is small, it provides enough context as to where the relevant objects are located.

Examples

It's important that users can clearly see that a part of an image has been magnified. Common implementations of visualizing zooming are symbolized magnifying lenses and zoom boxes.

Tip:
Some graphics programs already come with preset zoom effects that you can easily apply to an existing image or photo with a click of a button. In particular, this applies to screen capture tools.

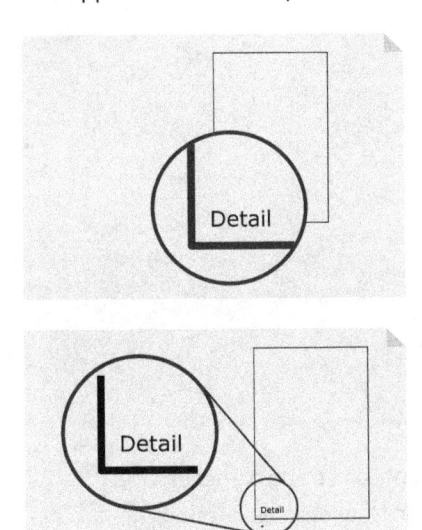

Images in general

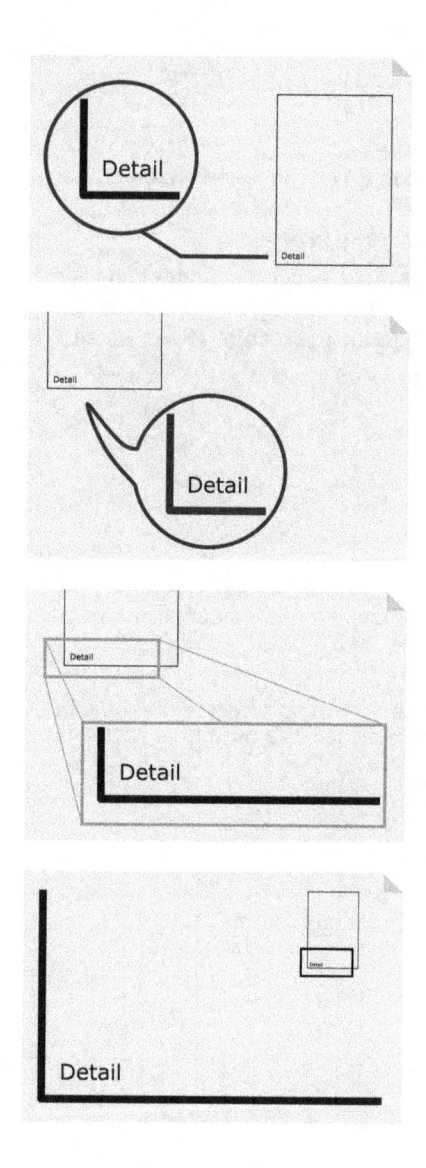

Detail

Detail

Detail

Detail

Detail

Detail

Detail

Detail

Detail

Tips for zooming

- To save space, make the zooming effect only as strong as necessary.
- Place the zoomed-in area as close to the original area as possible.
- In the zoomed-in area, use exactly the same perspective as in the original area.

Images in general

- Feel free to use multiple zoom areas in the same image. For example, you might want to point out several spots where something needs to be attached or connected.

Related rules
Guide the users' eyes 70
Emphasize what's important 77

2.3.11 Use text callouts or a legend for labeling?

A text *callout* is a short text label with a thin leader line that connects the text to the corresponding object in the image.

A *legend* is an explanatory text statement that's placed near the image but itself isn't connected to anything in the image. Instead, item numbers are used for identification. Either these item numbers can be printed directly on the corresponding objects in the image, or they can be positioned around the image and connected with leader lines similar to text callouts.

Both, text callouts and legends have their specific strengths and weaknesses.

- Text callouts mostly are more user-friendly (unless your images are very crowded and unless you to add long texts).
- Using a legend can save you time and money if your images need to be translated into many languages.

Example

Callout texts are part of the image:

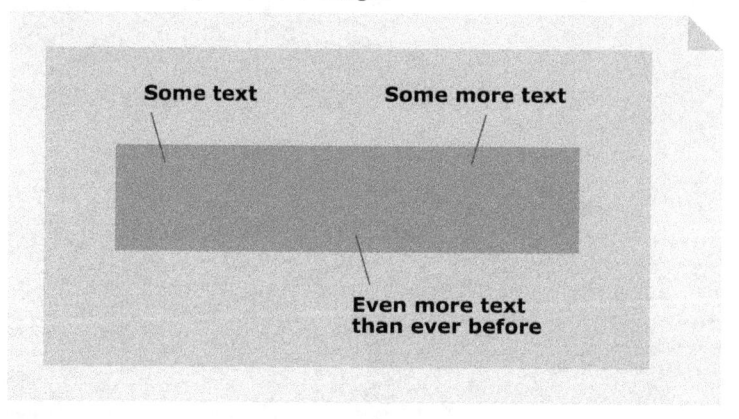

Some text

Some more text

Even more text than ever before

When a legend is used, only item numbers are added to the image. The legend goes below or next to the image and isn't part of the image itself:

Images in general

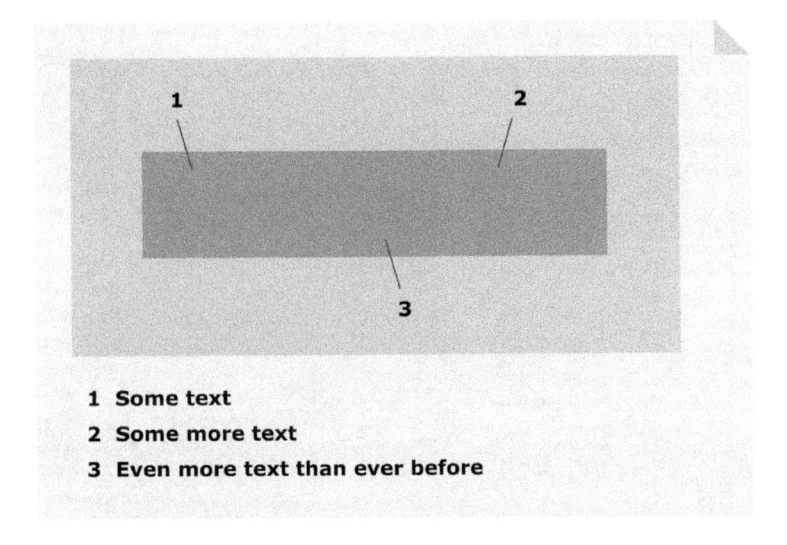

1 Some text
2 Some more text
3 Even more text than ever before

Advantages of direct labeling with text callouts

- The spatial proximity of the text to the part of the image that it relates to makes reading easy: Connections become immediately clear. Users don't need to waste any time looking for numbers.
- The image and the text can be viewed simultaneously, which facilitates understanding and improves memorability.
- It can't happen that in a printed manual an image ends up on another page than the explanation, or it can happen that in online documentation users need to scroll down to find the explanation.

Advantages of using a legend

- There's no risk of overloading the image with text. The image maintains its original simplicity and clarity. The visualization isn't obstructed.
- The graphical components of the image don't need to be made smaller to make room for the texts.
- The image doesn't need to be localized. The texts are part of the document's body text and can be translated along with the document, using the same translation workflow and translation tools.
- Legends even work with longer texts.

83

Images in general

> **Related rules**
> *Include text elements when necessary* 33
> *Tips on writing text within visuals* 37
> *Tips on formatting callouts and legends* 85

2.3.12 Tips on formatting callouts and legends

Text callouts, as well as the item numbers used for a legend, must be placed in the image in a way so that they don't obstruct the image. Often, they point to particular objects of an image with the help of leader lines, which add even more visual "noise."

- It must be easy for users to distinguish at a glance what belongs to the image itself and what are leader lines, labels, and legends.
- If you use item numbers and a legend, users must be able to find a particular number quickly.
- If you use leader lines, they must easily guide the eye to the right spot in the image. At the same time, they must not distract from the image.

Where to place the callout texts?

There are various places where you may put your callout texts. Each position has specific advantages and disadvantages.

It depends on the particular image which method works best. So you should decide on a case-by-case basis. As long as you use a common visual design for all the callouts in all images, it's OK if you vary the positions according to the characteristics of each particular image.

Pros and cons of placing the callout texts right **on top** of the graphical objects:

- Covers parts of the image.
- Minimal distance between an object and its description: No leader lines are needed, so extra clutter is avoided. The user's eyes don't need to move. Objects and their descriptions are perceived as a unit.
- No extra space is needed.

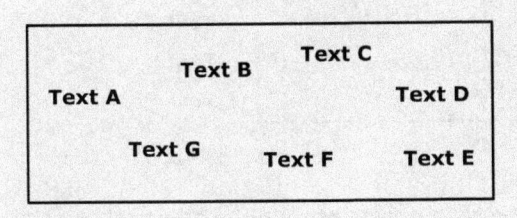

Images in general

Pros and cons of placing the callout texts **above and below** the graphical objects:

- Doesn't interfere with the graphical elements of the image.
- Adds to the height of the image. In printed documents, this can result in additional page breaks because a higher image is more likely not to fit on the current page than a less high image.
- If there are many callouts, there may not be enough space for all of them.
- Depending on the image, some of the lines may get quite long.

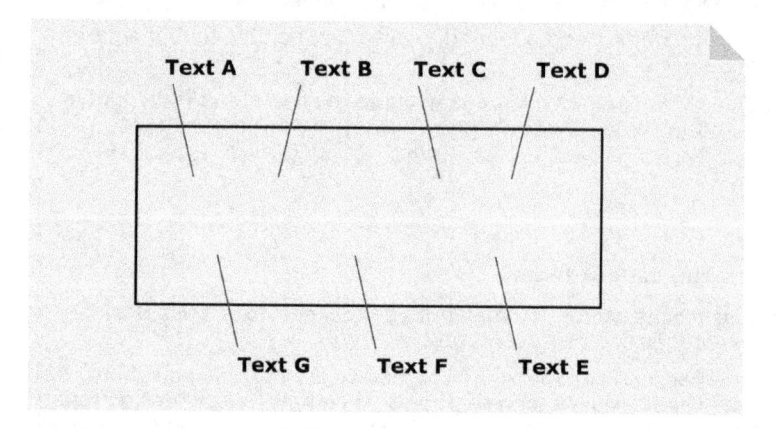

Pros and cons of placing the callout texts **left and right** to the graphical objects:

- Doesn't add to the height of the image, so pagination isn't affected.
- Adds to the total width of the image. So the main area of the image may get smaller.
- Depending on the image, some lines may get quite long.

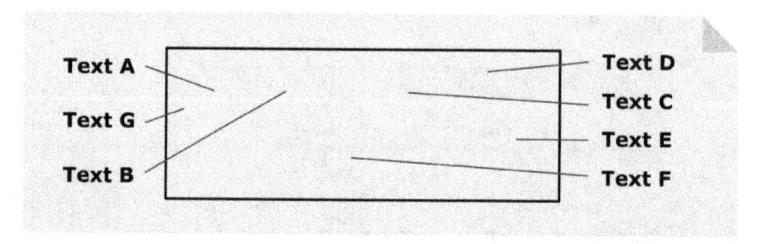

Pros and cons of placing the callout texts **around** the graphical objects:

- Keeps most of the leader lines short and thus avoids clutter.

86

- Provides more space even for longer texts.
- Adds to both the height and to the width of the image.

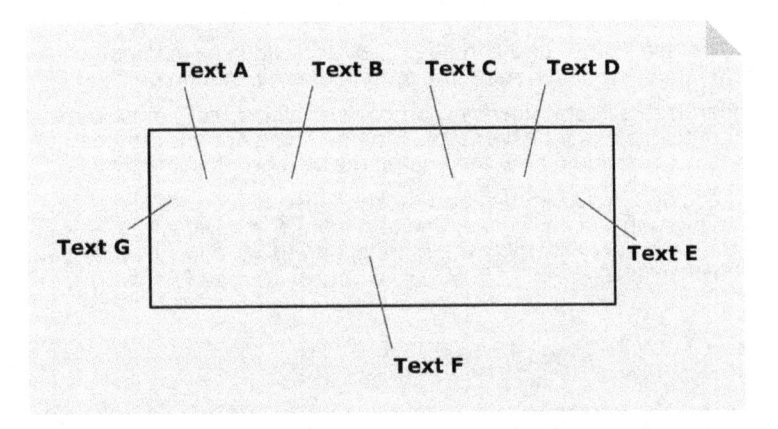

```
        Text A     Text B    Text C     Text D

Text G                                        Text E

                         Text F
```

General recommendations

No matter where you put the callouts:

- Place them so that the leader lines are as short as possible.
- Place them so that the leader lines intersect as few objects in the image as possible.

Where to put the legend?

If you have a legend, you can add it either to the image or to the body of your document.

- The advantage of having the legend in the *image* is the close spacial proximity. Both the image and the legend inevitably are on the same page and can be viewed at the same time.
- The advantages of having the legend in the *document body* are:
    - Translation is easier.
    - There are no space limitations.
    - In online documentation, the texts are searchable.
    - The texts may include cross-references or hyperlinks if needed.

Which order for item numbers?

To make it easy to find a particular item in an image, the items need to be numbered in a clear, systematic order.

Images in general

For numbering, use Arabic numerals. Don't use Roman numerals because many people can't read them properly. Alternatively, you may also use letters, such as *A, B, C*, or combinations of letters and numerals, such as *A-1, A-2, B-1, B-2*, etc.

Start near the upper left corner of the image. Then proceed in a systematic order. Users should be able to understand the system of organization quickly.

There's no standardized system. However, proceeding clockwise is most common and therefore typically the best choice. If the numbers are not arranged around the image but placed all over the image, the best order is line by line.

Depending on the image, it sometimes also makes sense to proceed along particular objects or in groups. For example, one assembly of a device is assigned the numbers *1* to *9*, the next assembly the numbers *10* to *19*, and so on. Or you create groups, such as [*A-1, A-2* ...], [*B-1, B-2* ...] or [*1A, 1B* ...], [*2A, 2B* ...].

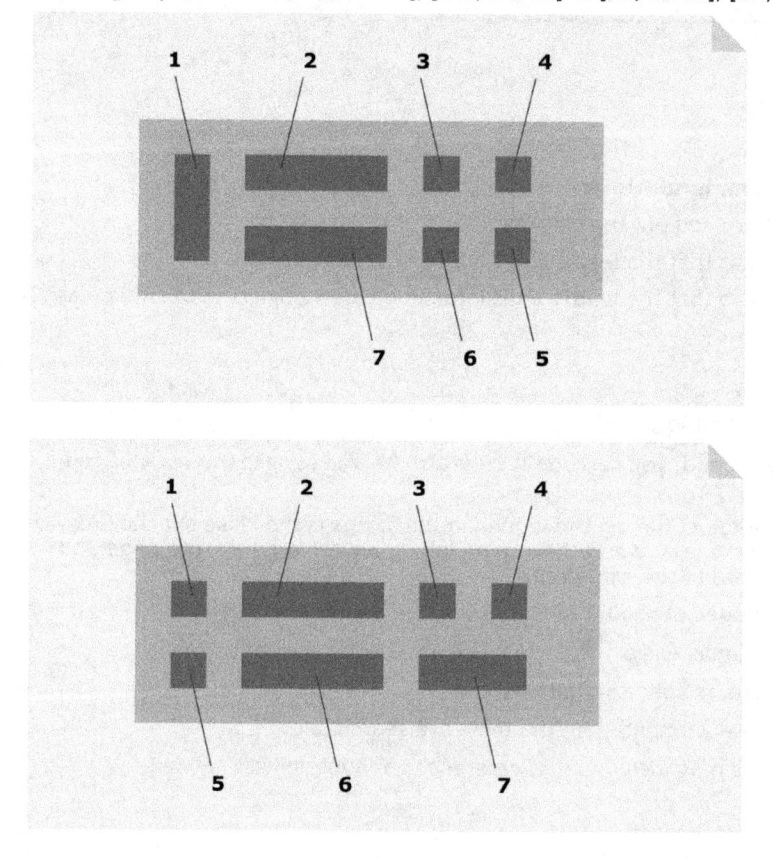

For the order of the numbers in the legend, look at how users will use the legend:

Only sort the list by number if users first look at an object in the *image* and then search the legend to **find the object's name or function**.

Example:

1	On / off switch
2	Program selection
3	Mute / Unmute
4	Brightness control
5	Contrast control
6	Timer
7	Volume control

However, if users know the name of an object and first look at the *legend* to **find the object in the image**, order the list alphabetically by object name.

Example:

Brightness control	4
Contrast control	5
Mute / Unmute	3
On / off switch	1
Program selection	2
Timer	6
Volume control	7

Tips for designing text callouts

If you only have a small number of callouts and if the callouts provide very important information, you can make them rather bold and colorful so that they really "call out" their message.

However, if you have a large number of callouts, rather minimalistic callouts are typically better. Not only do they have more visual appeal, they are also less obtrusive. So it's easier to recognize, interpret, and memorize the real image. In this case:

Images in general

- Avoid fancy callout bubbles and text boxes.
- Avoid background shading.
- Avoid very big font sizes and bold text.
- Avoid flashy colors.

This kind of callouts should not dominate but *supplement* the image.

✖ No:

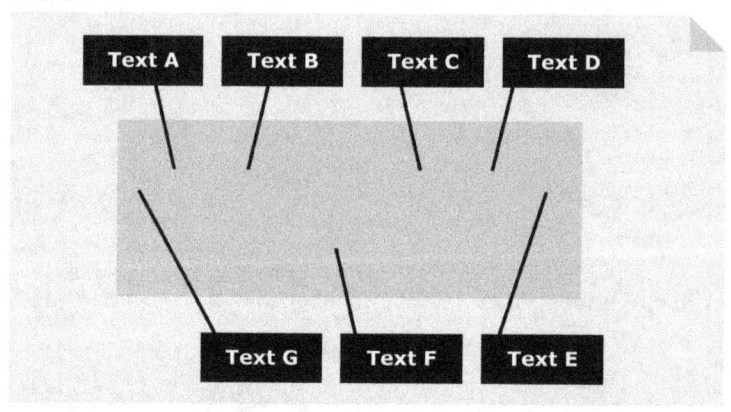

✔ Yes:

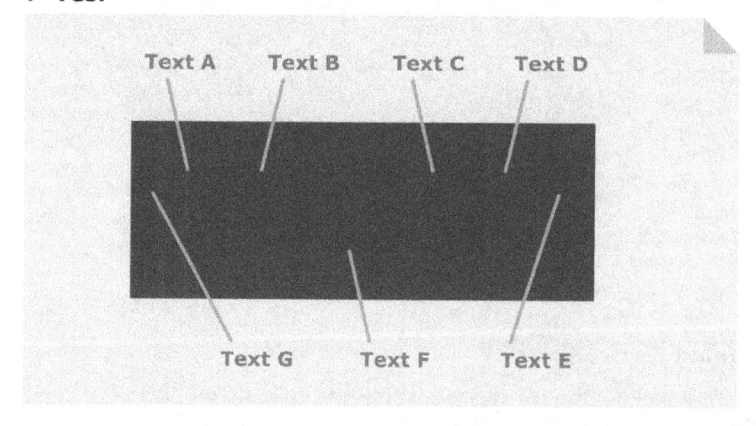

There's one important exception to this rule: If an image is very heavy with text, such as a screenshot of a software user interface, you need bolder callouts than with text-free images. For this reason, in screenshots visually rather dominating callout bubbles and boxes are quite common—especially if the callouts are positioned on top of the image itself but not around it.

Make sure that the callout text is visually distinct from other texts in the image.

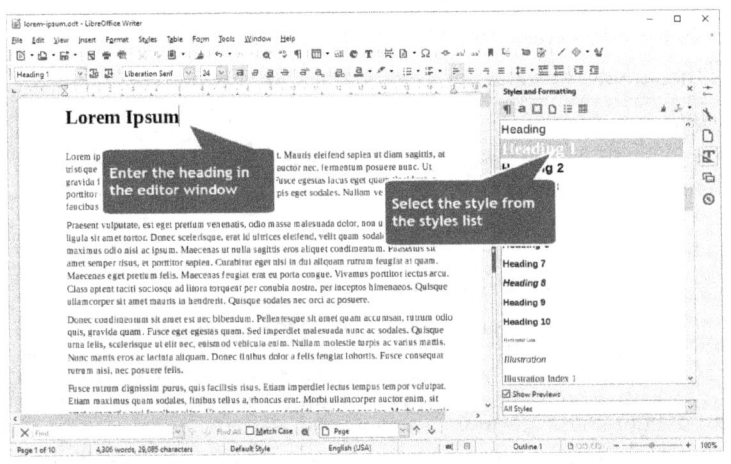

Tips for designing item numbers

If you use a legend with corresponding item numbers on the image, the numbers should be unobtrusive but yet easy to find.

You can best achieve this by choosing a design that's clear, simple, and somewhat different from the main image. In combination with the leader lines, the item numbers should look like an extra layer on the image, which users can then either focus on or look through.

Instead of adding a box or circle around each number for highlighting, better add some round or square monochrome background to each number.

If in the image, straight lines are dominating, use a round shape for the numbers' background.

If in the image round shapes are dominating, use a rectangular shape for the numbers' background.

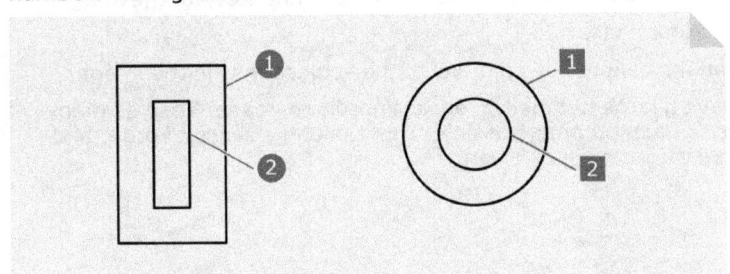

Tips for designing leader lines

The leader lines that connect callouts or item numbers to the image should:

- be unobtrusive
- be easily distinguishable from the lines of the image itself
- clearly guide the human eye

You can achieve this as follows:

- Use solid lines but no dashed or dotted ones.
- Use another line width than the one used in the image. Preferably use a narrow line width, but make sure that thin lines remain visible in case you scale down the image in your document. If you can't use a narrow line width, alternatively use a rather bold one but don't make the lines black. In this case, use a medium to light shade of gray, or use a soft spot color.
- Normally, don't add any arrowheads, dots, squares, or anything else at the lines' ends. In most cases, these elements aren't needed for clarity and just add clutter.
However, it must be clear what each callout line exactly points to. If you need to distinguish, you can use arrowheads to point to the edges of objects, dots to point to the interior of objects, and brackets to point to ranges of items.
- For slanted leader lines, you may add a horizontal shoulder line on which you later put the label text or item number.
Another variant is to keep the shoulder line short and to add the label text right or left to this line.
- Surround the leader lines with some white space. This prevents the lines from directly intersecting with lines in the image. Also, it ensures that the lines are still visible when they pass over image areas of the same color as the lines.
- Keep callout lines short.
- Try to route the callout lines over free or unimportant areas of the image.
- Don't cross callout lines.
- Avoid running the callout lines parallel to the subject lines in the image.
- When you have a large number of leader lines in an image: Keep as many lines parallel to each other as possible. This can add a sense of order and helps to make the image look clean.

Images in general

Some text callout

One more text callout

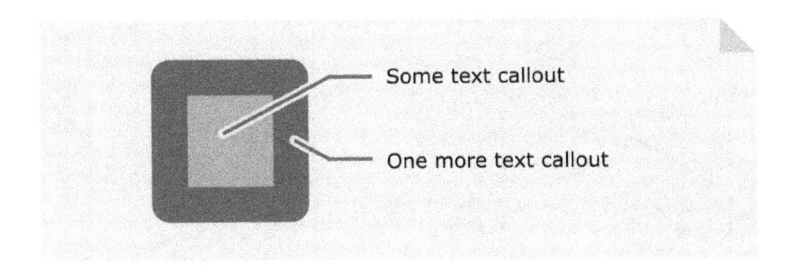

Exceptions in case the labels are more important than the image

Sometimes, actually the labels are the main entry points into an image rather than the graphical objects of the image. For example, users may already know the names of particular objects and are searching for the objects in the image. In a sales brochure or on a web site, the functions or benefits of a product are often the main message of an image.

Other than normal, in these cases make the labels more dominating than the image:

- Use large bold text or color for the labels or at least for the most important keywords within these labels.
- Use leader lines that are bolder and darker than the lines in the image.
- Add arrowheads or dots to the leader lines' ends.

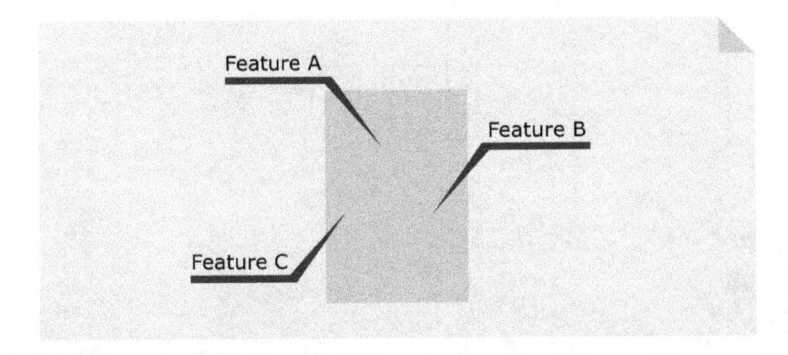

Feature A

Feature B

Feature C

Dimension lines

Don't let dimension lines clutter an image.

Place the dimension lines and the corresponding numerals at the sides, like it's common in construction plans. Do so even for non-technical audiences. However, for better readability always keep the text direction horizontal.

Images in general

✘ No:

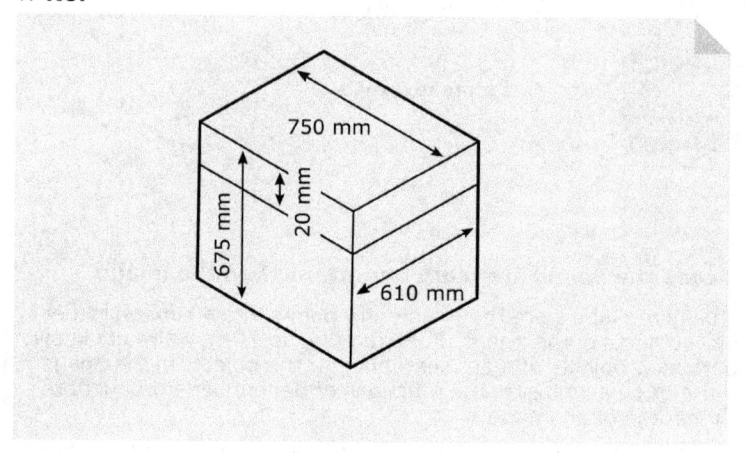

✔ Yes:

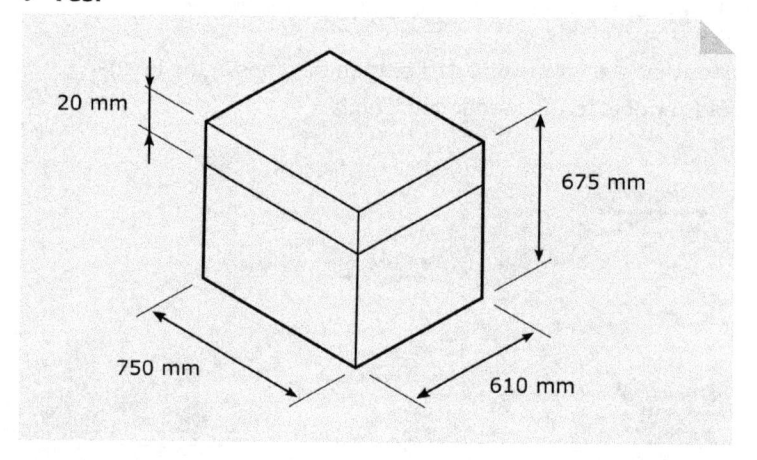

Related rules

Include text elements when necessary 33
Use text callouts or a legend for labeling? 82

94

2.3.13 Tips on formatting image texts in general

The texts in images are essentially a part of your *written* communication. Thus, the same principles apply for choosing fonts as for the body text of your document.

- Readability: The text should be as easy to read as possible.

    Use a simple font that the users are familiar with, but don't use anything fancy. Many people feel that sans-serif fonts are easier to read than serif fonts. Sans-serif fonts have clearer lines and thus produce less "visual noise", which is very important in complex images for technical documentation.

- Consistency: All images should use a common set of design elements. This includes a common font.

    Best use a font of the same font family that you also use for the body text of your document.

- An additional requirement is that there's usually not much space available for the texts. So the spacing of the font should be rather small.

Tip:
A good combination can be to use the same font for image texts that you use in the document for the column headings of tables or for topic headings and subheadings. These elements also need small font spacing. Many font families include a condensed variant of the given font. So you can use the base font for body text and the condensed variant for headings and image texts.

Font style

Keep it simple and stupid. Don't use many different font styles. Normally, use only one font without any additional formatting.

In some cases, you may like to have one bolder and one lighter formatting in order to be able to guide the users' eyes through longer text sections. For example, in an image that outlines the parts of a device, you could use the bolder variant for the parts' names (eye-catcher and key information) and the lighter variant for some extra description on what each part does.

Some fonts are available in special bold variants.

Special light variant of the font
Base variant of the font
Base font variant of the font put bold
Special bold variant of the font

Font size

To save space, you can choose a font size that's slightly smaller than the body text of your document. Because in images texts are short, this doesn't affect the readability as badly as it would in the body text. However, make sure that the letters and numbers are still large enough so that even people who have a poor eyesight can read them. The text must be *legible*, not just *recognizable*.

Setting an exact font size can be tricky:

In some authoring tools, you can embed an image and then add the image text on top of the image in the authoring tool. In this case, you have full control over the font size.

However, if you add an image that has already text included in the image itself, the final font size in your document depends on how you scale the embedded image. For example, even if you consistently use a font size of 12 pt in each image, when you scale one image to 100% and another to 80%, in your document the font sizes will be different: 12 pt for the first image but only 80% of 12 pt (that is 9.6 pt) for the second image. To account for the scaling, you'd need to adapt the font size in the image, which is not practical.

In practice, you can either:

- refrain from scaling images
- define a limited number of possible scaling factors, then for each image decide which of these scaling factors you're going to use, and then set the font size in the image accordingly
- just live with the fact that the text size isn't exactly identical in all images

Which method works best for you essentially depends on the time and budget that you can spend, as well as on how much your images differ from each other in size. However, typically most users will never notice any difference in font size, and even if they do, they won't mind.

Line spacing

To save space, also the line spacing can be a bit smaller than in the body text of your document. Because image texts are short, this doesn't have any important impact on readability.

Left and right alignment

You have the following options:

- You can left align all texts.

  This design provides the best readability if the individual texts span multiple lines.

- You can left align all texts that are in the left half of the image and right align all texts that are in the right half of the image.

  This design looks most harmonious and symmetrical, especially if your document's body uses justified text.

Horizontal and vertical alignment

Always use horizontal text. Do not use vertical or rotated text. Also, do not stack the letters on top of each other. All of these alignments are very hard to read and therefore make it difficult to concentrate on the content.

✖ **No:**

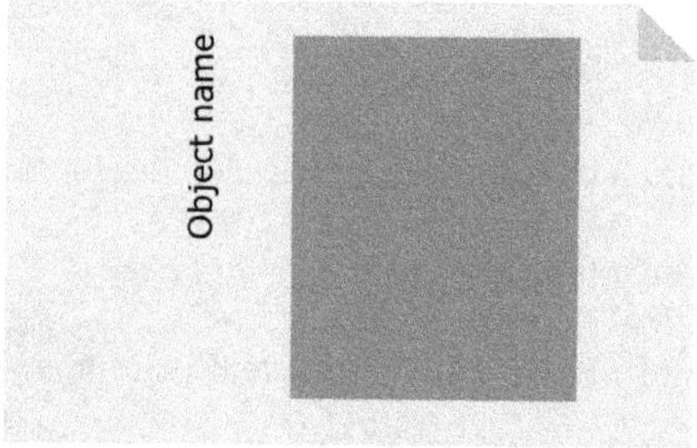
Object name

✖ **No:**

O
b
j
e
c
t
n
a
m
e

✔ **Yes:**

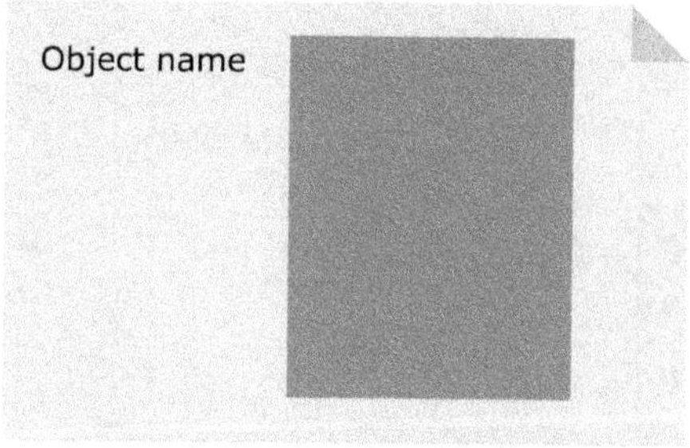
Object name

Put multiple texts on the same line to add a sense of order and to make the image look clean:

Images in general

✘ No:

[Image showing a rectangle with cramped, overlapping "Text" labels above and below]

✔ Yes:

[Image showing a rectangle with evenly spaced "Text" labels above and below]

✔ Yes:

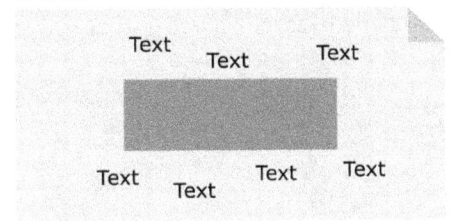

[Image showing a rectangle with "Text" labels above, and "Text / Some longer text" and "Text / Also longer text" below]

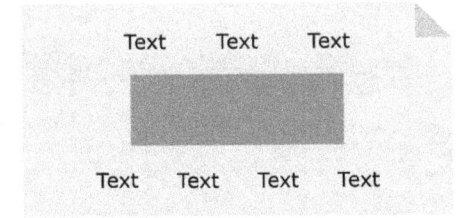

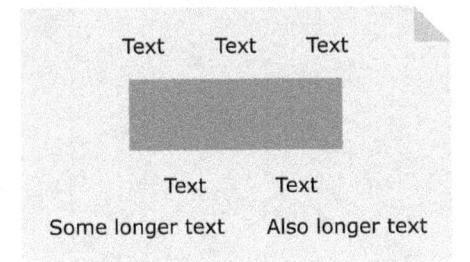

Avoiding ragged, pixelated letters and leader lines

When you add text to a raster image (such as a photo or screenshot) and then enlarge this image in your document, letters and leader lines may look ragged. This happens because there are only a limited number of pixels saved in the raster image, which now become apparent as a result of the magnification.

To prevent this problem, you have the following options:

- You can save the image, including the text, as a vector image (such as SVG) and embed the vector image into your document. This is the best solution if your tool set supports it. The pixel image in the background remains the same, but as the text is vector-based, it can be scaled to any size without becoming blurry or pixelated.

99

Images in general

- You can enlarge the image before adding the text, add the text, embed the image into the document, and then scale the image down *within the document* as required. This increases the size of the image file, but because there are more pixels available, the letters and lines look much crisper.

> **Related rules**
> *Include text elements when necessary* 33
> *Tips on writing text within visuals* 37

2.3.14 You rarely need figure titles

Many authors of technical documentation add a title above or below an image, which briefly summarizes what the image is about.

In most cases, you *don't* need these titles. Typically, they are nothing but useless clutter on the page.

You *only* need figure titles:

- if you aren't confident that your images are designed well enough to be clear in what they are showing (better revise the images then)
- if you aren't confident that the document text and its headings add enough information to make clear what the images are about (better revise the text then)
- if you want to be able to create a list of figures (and have a good reason why users need a list of figures)

If you have designed your images well, it's mostly obvious what they are showing

A good image should be so clear that it's obvious from looking at it what it's about. If this isn't the case, probably something is wrong with the image.

There may be some exceptions when an image shows an isolated object or when an image shows a complex diagram that can't be grasped at a glance. In a case like this:

- You can introduce the image with a short phrase in the body text.
Example: "The following image shows ...."
- You can add some short descriptive text to the image itself, best placed near the upper left corner of the image.
Examples: "Power unit" or "Recharging time depending on temperature"

If you have placed your images well, it becomes clear from the context what they are showing

In good documentation, it should be obvious from the body text what each image is about.

If you place your images precisely in the text where they fit into the content, and if your text is clearly structured and has enough headings and subheadings, there's no need for figure titles.

See also *Position each image precisely* 51

Chances are you won't need a list of figures

An argument sometimes used in favor of having figure titles is that you need them in order to be able to create a list of figures.

Only add a list of figures if this list actually adds some real value to your document. Typically, having a list of figures only makes sense if users want to look up specific images directly, for example, specific flow charts or wiring diagrams.

Don't add a list of figures only because your authoring tool can create one automatically. It may be tempting to use such an automatic function to produce a few nice-looking pages in no time, but in most user manuals a list of figures simply doesn't make sense but just adds to the page count.

Often, a well-designed table of contents can do a much better job in guiding users towards the right images than a list of figures can.

Tips for phrasing figure titles in case you have decided to use them

If you *do* add a figure title, it should communicate:

- What are the users looking at?

  This is important for orientation.

- If it isn't evident: Why are you showing the image?

  This can be the key to understanding an image and for looking at the right details.

- Optional: Limitations

  This can help users to skip an image if it's not relevant in their case.

Figure titles need to be short and to the point; a line and a half is the maximum.

In English, prefer sentence-style capitalization for figure titles.

Don't end a figure title with punctuation even if the figure title is a complete sentence.

Don't begin a figure title with any introductory word or phrase, such as "Shows ..." or "In this image you can see ...."

Other than in the body text of the document, don't give any instructions. Don't talk to the user directly.

Try to put the keywords in the figure title as far to the beginning as possible. This makes it easier to skim over the figure titles. If users find from the first words of a figure title that the image doesn't show what they are looking for, they can immediately continue somewhere else.

✘ **No:** *Here you can set the speed*

✖ **No:** *Set the speed here*
✖ **No:** *Shows the speed setting*
✔ **Yes:** *Speed setting*
✔ **Yes:** *Speed setting on the control panel*

✖ **No:** *Cables*
✖ **No:** *Remove cables safely*
✔ **Yes:** *Safe removal of cables*
✔ **Yes:** *Safe removal of cables (Model C only)*

Tips on formatting figure titles

In case you *do* decide to use figure titles, you should format them so that they are both:

- not too dominating on the document page
- quickly findable if users are actually looking for them

Tip:
Often, a good formatting is to combine a bold character style with a slightly smaller font size than the one used for the body text.

Related rules

Page layout of pages with images 58
You rarely need figure numbers 104

2.3.15 You rarely need figure numbers

Many authors of technical documentation add a number to each image to be able to later refer to the image in the text (example: "… see image 7").

Often, this isn't the best solution because:

- The numbers (often combined with a figure title) make the text longer but don't add any information on the subject.
- If you need to refer to an image instead of including the image right where it's needed, following the referral means some extra work for users. They need to leave their current position in the text, search for the image, and, after looking at the image, again search for the place where they left off.
- In case you're creating online documentation, figure numbers just don't make much sense at all. True online documentation is hypertext where there isn't any fixed sequence of topics similar to the fixed sequence of chapters in a book. So you can't number images from "start" to "finish." You could only number them *within* a topic—ending up with having multiple "image 1", "image 2", etc. in your documentation.

If you've placed your images well, you don't need to refer to them in the text

If you place an image well right where it fits into the content (see *Position each image precisely* 51), you typically don't need to explicitly refer to the image at all. If it's naturally clear which image belongs to a text section, you don't need any extra references such as "see image 7." You can just talk of "the image" and it's evident which image you mean.

Sometimes, you need to refer to an image that's not in the same topic but somewhere else. In this case, the best solution is to add the image a second time. Even though this involves some degree of redundancy, it has several advantages:

- Users don't need to follow a cross-reference or click a link. This saves users time and it doesn't take them away from the primary context.
- The chapter or topic remains fully self-sufficient, which is a key requirement in single source publishing. If, for whatever reason, the chapter or topic with the "original" image doesn't exist in a particular version of the documentation, the second one remains fully understandable and doesn't have a broken link.

Don't mind the extra space that the second copy of the image needs. This is a good investment into the usability of your documentation. Also, it only affects printed documentation. In online documentation, you can repeat each image as often as you like. Nobody will notice, let alone bother about it.

An option can be to refer to the topic that contains an image rather than to the image itself.

Using figure numbers to overcome poor page breaks

In printed or PDF manuals, it sometimes happens that an image doesn't fit onto the rest of a page. So there's an automatic page break, leaving some whitespace. The problem with this isn't so much the whitespace but the fact that users may not be aware that the current topic actually is not yet finished but continues on the next page.

If you want to make your manual perfect and can invest some extra time, you may like to include a note that points to the image, such as "see Figure 7 on page 55." However, even in this case, you don't necessarily need figure numbers. You can just as well phrase the note "see the next image." In fact, this variant is even more user-friendly because it avoids the information about the number and the page, which is not really needed here.

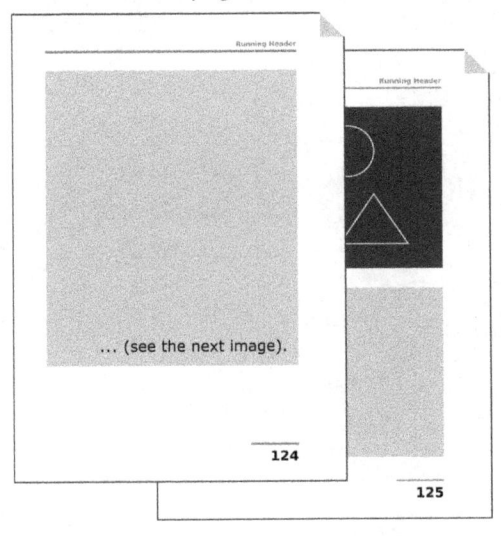

Note:
If you also create online documentation from the same source, you don't need these references there. Most professional authoring tools enable you to make text conditional so that it doesn't show up in a particular medium.

Tips on formatting figure numbers

In case you *do* decide to use figure numbers, you should format them so that:

- they are not too dominating on the page
- yet they can easily be spotted if users are actually looking for them

Images in general

Tip:
Often, a good setting is to combine a bold character style with a slightly smaller font size than the one used for the body text.

The numbers themselves can either:
- be consecutive throughout the whole book
- consist of the main chapter number followed by a hyphen plus a consecutive number that restarts with each main chapter

Examples:

✔ **Yes:** *Image 7*

✔ **Yes:** *Image 4-2*

> **Related rules**
> *Page layout of pages with images* 55
> *You rarely need figure titles* 104

2.3.16   Color or black and white?

Unless you ship your documentation exclusively printed on paper in black and white, you need to decide whether your images use color.

- With **photos**, both color images and black-and-white images each have their particular strengths and weaknesses. Often, it's a good solution to combine black-and-white with one spot color for highlighting.
- With **screenshots**, use full color if you can.
- With **line drawings**, in most cases it's best to use only very few colors or even just black and white plus one spot color.
- With **diagrams**, in most cases it's best to use multiple colors, but only as many as necessary. Subtle colors work better than bold ones.

> **ⓘ Important:** When using any colors in documents that you supply electronically, make sure that no information gets lost in case users print the document on a black and white printer.

Color with photos

When a device is shown in full color, it looks most realistic. Therefore, users can identify it most easily.

When a device is shown in black and white or as a grayscale image, you can reserve color as a strong means for highlighting particular objects within the image.

Thus, it depends on the purpose of your images what works best:

- If you want to show a device as a whole, best use color.
- If you want to show the location of a particular component or control on the device, consider using a grayscale image in which you add some color only to the particular component or control that you want to highlight.

    In doing so, you can either:
    - Apply the original full image colors on the highlighted area, which can look a bit obtrusive because the contrasts to the black-and-white parts of the image are very harsh.
    - Colorize the relevant object in the image with only one spot color. If you use a light shade of your product logo's main color, for example, this can look very elegant and professional.

Images in general

- Add a colored line around the relevant object in the image, but leave the object itself black-and-white. This is a good option especially if the highlighted object has a very characteristic shape.

Another variation is to reduce the color saturation outside the relevant object instead of using grayscale.

Color with screenshots

Screenshots are typically shown in full color.

This is a good practice because other than physical objects, a software user interface doesn't have any three-dimensional characteristics. So it could be quite hard to identify an element of a user interface on a black-and-white image quickly.

Color with line drawings

Line drawings by nature aren't exactly realistic. Even more so, typically they are a purposeful simplification of reality. Using black and white instead of colors can do a good job in adding to this simplicity.

Similar to photos, it can be a good option to combine a black-and-white drawing with a single spot color for highlighting. When you use the color only on particular objects, this can put a strong emphasis on these objects and guides the users' view to the right places in the image.

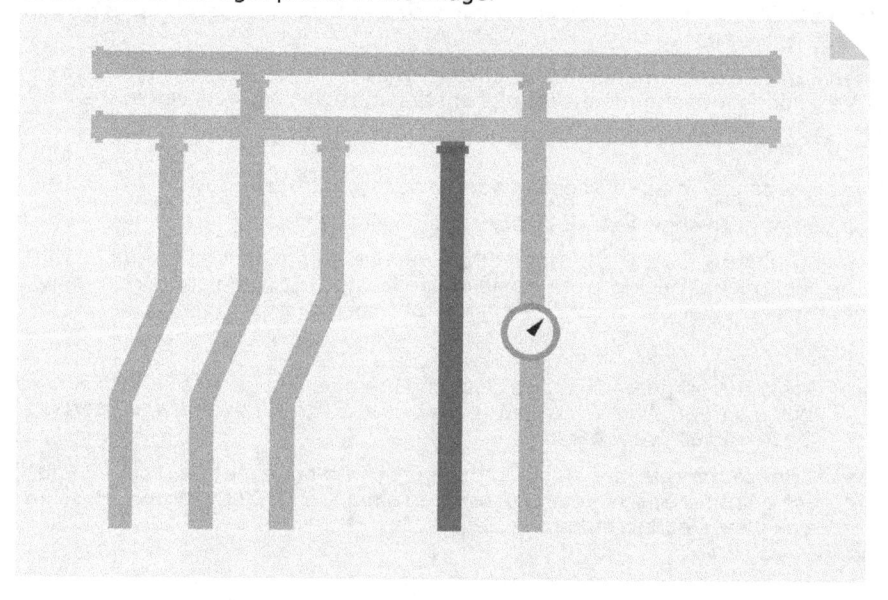

Color with diagrams

Many diagrams greatly benefit from using color as a means of distinguishing different kinds of data.

So if you can use color, do so.

However, prefer colors that are rather subtle and unobtrusive. Color should be a supporting tool, but it should not dominate the image visually.

Optimizing colors for printing on a black and white printer

When supplying any documents electronically, you should be aware of the fact that some users may print these documents (or parts of them) on a black and white printer.

So make sure that when your documents are printed in black and white no information is lost:

- When choosing colors that have a particular meaning, select colors that clearly vary in contrast and thus produce a different shade of gray on the printout.
- If there's no very distinct difference in contrast, add a second coding.

Example:

Imagine you're using solid red lines in your images for pointing out some special elements. When printing the images on a black and white printer, these lines are almost just as black as the black lines in your image. So the highlighting effect is completely lost.

You could now use, for instance, orange instead of red, which would produce a shade of gray on the printouts instead of black. But the color orange may not blend nicely with the design of your brand, so you want to stick with red.

An elegant solution in a case like this can be to use a lighter shade of the original color plus at the same time increase the line width. When printed in color or shown on a computer screen, the total amount of red in the image remains about the same. So the overall visual impression essentially doesn't change very much. However, when printed in black and white, the lines now much better stand out from the rest of the image: They are wider and thus bolder, and they have a clearly distinct color (shade of gray). Yet the lines aren't dominating the image too heavily.

Converting color images into grayscale images

Depending on the image editing software that you use, there are big differences in the quality of how images are converted from color to grayscale. Experimenting with the settings and perhaps even experimenting with different programs can be well worth the effort.

Images in general

Select a conversion method that results in a black and white image in which all colors are clearly distinguishable.

Depending on the colors in your images, different images may need different settings.

Related rules

Use color sparingly 25
Which color mode? 116

2.3.17 Which resolution?

The more pixels a raster image has and the closer these pixels are printed together, the higher is the image's resolution and thus the visually perceived quality.

The resolution is measured in DPI (dots per inch). It's always a result of two factors: the number of pixels and the print size of the image.

The downside of having images that have a large number of pixels (and thus potentially a high resolution) is that these images also have a large file size. Because images are two-dimensional, the file size increases quadratically with the number of pixels in each direction. Thus, finding the optimum image resolution is always a compromise between quality on the one hand and performance and file size on the other.

Typically, good trade-offs for the resolution are:

- 150 to 200 DPI for standard printing on an office laser printer (so you need about 800 pixels in width for about 10 cm)
- 300 DPI for quality offset printing (so you need about 1200 pixels in width for about 10 cm)

Important: It doesn't make any sense to use a resolution higher than the one that the used printer can print. The images would then be automatically reduced to this maximum resolution.

In online documentation, the DPI settings stored in an image don't have *any* effect. If an image has more pixels, it just becomes larger.

Understanding vectors and pixels

When creating diagrams and infographics, it's best to create a **vector image** (see also *Which file format?* 118). Vector images can be enlarged to any size without becoming blurry or pixelated. This is because the images are rendered only at runtime. When you draw a line, for instance, the image file only stores the coordinates of the line's starting point and ending point. When the image is displayed on screen or printed on paper, the line is drawn between the two points, using the full resolution of the device. So the line appears crystal clear no matter how much you zoom in.

When taking photos and screenshots, these images are **raster images**, consisting of a finite number of pixels. If you zoom in on a line in a raster image, eventually you'll see the square pixels that the line is composed of. The line then looks jagged and pixelated. Given a particular image size, the more pixels the image consists of, the more you can zoom in before you see the pixels.

Images in general

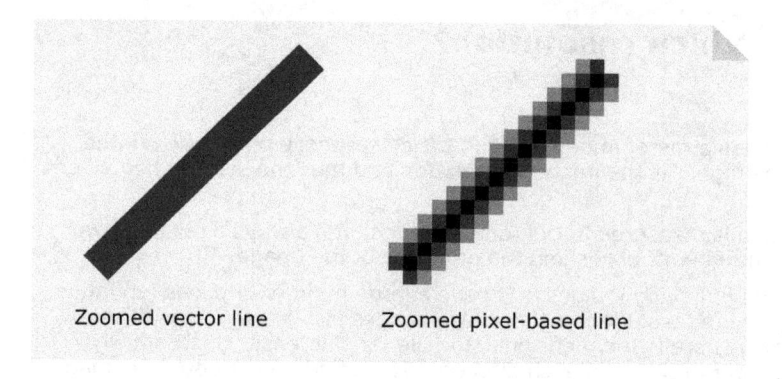

Zoomed vector line   Zoomed pixel-based line

Understanding DPI and resolution

DPI is a measure for the resolution and stands for Dots Per Inch.

If an image has many dots per inch, it looks sharp. If an image has only a small number of dots per inch, however, it looks blurry or pixelated.

An image with many dots per inch can be made very large, or you can zoom in on a very small detail. The quality then still remains good. However, an image with a small number of dots per inch only looks sharp when it's printed or displayed in a small size. When you zoom in on such an image, you'll see all the pixels.

DPI are not a property of an image

It's important to understand that the DPI value essentially is *not* a property of an image itself even if it can be specified for an image and saved along with the image in the image file. The final resolution or DPI results from the number of pixels *in combination with* the size in which the image is printed.

Example: Screenshots typically have a default resolution of 96 DPI. If you increase the DPI to 200 or 300, the image shrinks to about two thirds when printed on paper. This happens because the image has only a limited number of pixels. To produce a high DPI value, these pixels need to be packed into a smaller space. So the image gets smaller.

Original screenshot at **96 DPI**

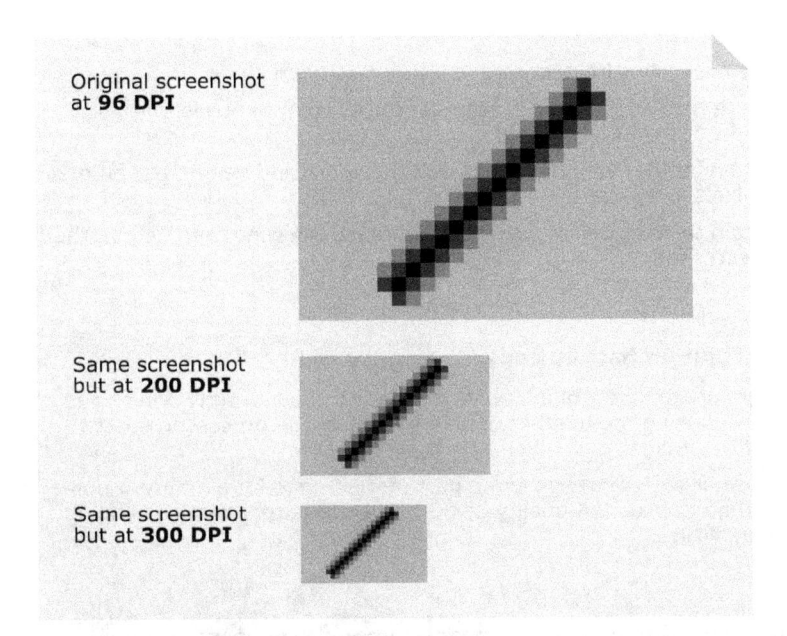

Same screenshot but at **200 DPI**

Same screenshot but at **300 DPI**

The DPI value stored with the image can be seen as a command as to how crisp you want the image to be when it's included in your document. In the example, if you tell the screenshot that it has a DPI value of 300 instead of 96, when you embed the image into a document in your word processor, the word processor makes the image so small that when printed on paper, 300 DPI will be achieved.

If you later resize the image in your word processor, the DPI on paper will change. For instance, if you double the width of the image, you'll end up with only half of the DPI—in this case, 150 DPI.

When using an image in online documentation, any DPI setting stored with the image is typically ignored. Unless you resize the image explicitly, one pixel of the image becomes one pixel on screen. The actual resolution then depends on the resolution of the user's particular monitor. (Note: Depending on the browser or viewing application used, on high-resolution displays sometimes one pixel in an image uses multiple pixels on screen. This reduces the resolution but doesn't change the principle.)

DPI when exporting vector images

When you export a vector image to a raster image, most programs allow you to specify the following properties:

- *DPI*
- *number of pixels*
- *width* and *height* in mm or inches

However, these properties influence each other. For example:

- If you increase the DPI value and maintain the number of pixels, the width and the height decrease.
- If you maintain the DPI value and increase the width and height, the number of needed pixels increases.
- If you maintain the number of pixels and increase the width and height, the DPI value decreases.

Keep a full-resolution backup copy

When shooting a photo, use a high resolution even if you currently don't need it. Then, before reducing the number of pixels in an image processor, save a backup copy of the image.

If you later find that you can reuse some part of the image for a different purpose, you can then use the full-quality backup to enlarge that part and still obtain sufficient quality.

Best enlarge raster images with a special image processor

With screenshots, the total number of pixels is given by the captured window size and cannot be changed. Thus, if you don't want a screenshot to become too small when embedded into a document, you may need to work with a lower DPI value than for photos. In most cases, this is OK and produces sufficiently good quality.

If a screenshot or photo doesn't have enough pixels but you need to enlarge it to a large extent: Instead of just resizing the image in your authoring tool you may obtain better results if you first enlarge the image in a professional image editor and then embed this enlarged version of the image into your document. There are even some very specific programs and plugins available for enlarging low-resolution images. The algorithms do not just enlarge the existing pixels within the image but add additional pixels by means of vectorization and artificial intelligence.

Best downsize raster images in your authoring tool

Unless your original images are very large, it's often most efficient to downsize them in your authoring tool rather than in an external image processor. Most authoring tools then automatically produce smaller versions of the files for online documentation rather than just setting the HTML image width and image height properties. So the initial file size doesn't matter.

This approach has several advantages:

- The original files are always saved with the project. So you don't need to keep extra full-resolution backup copies.

Images in general

- You can easily change the size at any time.
- If you produce both online documentation and printed documentation from the same source, the printed documents can automatically use the full-resolution versions of the images. So you don't need to handle two different versions of each image separately.

Related rules

Which color mode? 116
Which file format? 118

2.3.18 Which color mode?

> Advanced image processors let you choose whether an image uses RGB color mode or CYMK color mode.
>
> - RGB stands for the three primary colors used by computer displays: Red, Green, and Blue. RGB is primarily intended for electronic content.
> - CYMK stands for the four primary colors used in printing on paper: Cyan, Yellow, Magenta, and "Key" (black). CMYK is typically required for professional printing.

Which mode to use for print, online, and single source publishing

- If you're going to create and ship your documentation **only electronically** (HTML or PDF), use RGB.
- If you're going to ship your documentation **only on paper**: Most likely your print shop will ask for CYMK. If you send them RGB images, the colors on paper may look quite different from the colors on your computer screen.
- If you're going to use your images **both for online documentation and for documentation printed and shipped on paper** (single source publishing), you have the following options:
    - You can use only RGB.

        In case your documentation is printed in black and white, or if colors on paper don't need to be completely realistic, this should be fine.
    - You can use RGB but have the images converted to CYMK before sending your documents to the print shop. Some tools can do this automatically.

        This is a good option if efficiency is more important than perfection.
    - You can use different versions of the images. Most authoring tools for technical documentation enable you to use conditional content. So you can embed an RGB version of an image and tell it only to appear when creating online documentation. In addition, you can embed a CYMK version of the same image and tell it only to appear when creating printed documentation.

        This process lets you tweak each image individually and is best if you need very realistic colors and don't want to rely on an automatic conversion process.

Additional tips

- Switching the color mode may slightly change the visual appearance of the colors. So set the color space *before* editing an image, not afterwards.

- With photos, change as few color settings as possible. Each change may result in a slight loss of quality.
- In case you're planning to ship your documentation on paper: Contact your print shop early to discuss their particular requirements. Not all print shops have the same requirements. Knowing these early can save you the time for later converting the images.
- No matter who is going to print a document, a professional print shop or the users on their own printers: Be aware of the fact that the colors will most likely not look exactly like the colors on your own computer screen.

Related rules

Color or black and white? 107
Which file format? 118
Which resolution? 111

2.3.19 Which file format?

> Which file format works best, depends on the particular type of image. So you might need to use several formats within the same document.
>
> Basically, you're free to choose any format that your authoring tool supports. However, if you're creating online documentation, it's best to use a format that can be viewed directly in a browser and thus doesn't need any conversion.
>
> Also, it's always best to use a common format rather than a very specific one for various reasons:
>
> - You might like to switch to a different image editor one day.
> - You might like to switch to a different authoring tool one day.
> - You might like to use the same images also elsewhere. For example, for a marketing brochure, on a web site, or in a presentation.
> - You might need to send the images to other persons for reviewing, editing, or translation.

Photos

For photos, typically JPEG is the best choice. JPEG is produced by most digital cameras, and it can be displayed by all web browsers. So it doesn't need any conversion. For most photos, JPEG provides good compression. Therefore, the image files are rather small. If you need especially small files, you can apply lossy compression, which makes the images lose some of their crispness.

Avoid using PNG for photos. In most cases, it produces very large files.

Avoid using GIF for photos either. It only supports 256 colors, which results in very poor image quality with photos.

Screenshots

For simple screenshots, PNG works best. It yields good, lossless compression with images that have large areas of the same color, like it's typical with software user interfaces. Make sure to use PNG 24. PNG 8 can use only 256 colors.

If you edit your screenshots, especially if you add callouts or other texts, SVG is a good choice. Similar to PNG, SVG can be directly used in HTML-based online documentation. Additional advantages of SVG are:

- SVG images may be styled with CSS (Cascading Style Sheets) and even animated.
- Texts in the images are searchable.

- Texts in the images can be read by screenreaders, making your documentation more accessible.
- Texts in the images can easily be translated with a simple text editor or with the help of a translation memory system.

Don't use JPEG for screenshots. For screenshots, lossy JPEG compression doesn't work well and results in very poor quality.

Don't use GIF for screenshots either. Because GIF can only use 256 colors, it's not appropriate for modern user interfaces. On a GIF image, a user interface may look completely different than it looks on screen. Soft gradients may appear as harsh steps.

Diagrams and infographics

For all sorts of diagrams and infographics, best use a vector-based image format. Vector-based images can be enlarged at any time without becoming blurry or pixelated. Also, text and all other objects within the image remain fully editable. So you can easily update vector images at any time.

If you can, use SVG. Compared to other common formats, such as EPS, it has some major advantages:

- SVG images can be displayed by web browsers. Thus, SVG images don't need to be converted into a pixel-based format when using the image for online documentation.
- In online documentation, SVG images can be styled with CSS and even animated. This makes it possible, for example, to disclose content progressively.
- In online documentation, texts in the images are searchable.
- In online documentation, texts in the image may be read by screenreaders so that the images provide optimum accessibility.
- Texts in the images can easily be translated with a simple text editor or with the help of a translation memory system.

Animated images

For simple animated images, best use PNG. Other than GIF, which also supports animation, PNG images can use more than 256 colors.

For more complex and interactive animations, SVG is a good choice.

Icons and symbols

For small icons, PNG mostly is the best file format. If there are only a few different colors in the icon, GIF may produce slightly smaller files. However, as the

Images in general

files are very small anyway, this doesn't really matter.

For monochrome symbols that need to be scaled to different sizes and that are needed in many places, an alternative to adding these symbols as an image can be to create a custom image font. For example, you might like to create such a font for referring to special keys of a product.

Tip:
In online documentation, you can embed the font with the CSS command `@font-face`. Alternatively, some authoring tools can convert symbol fonts to small vector images for publication.

Compression when converting to PDF

When creating a PDF file from a document, pay particular attention to the compression settings for images.

- Only if a small file size of your document is very important, enable lossy compression. Even if the images in the PDF look OK on screen, they may look blurred or pixelated when printed on paper. Make a printout to check the quality.
- Choosing *loss-free* compression is uncritical but typically produces larger files than lossy compression.

Transparency

When creating online documentation, don't underestimate the importance of a format's support for transparency. You need a transparent image background more often than you might think. File formats that support transparency are PNG, GIF, and SVG.

Using a monochrome background instead of a transparent one may be a workaround, but it has some major downsides:

- If you have a monochrome background, its color must be the same as your document's background color. In case you later decide to change the design of your document, you need to edit all images accordingly.
- You can use the same image also for creating printed documentation (or PDF) only if the images's background is white.

Related rules

Which resolution? 111
Which color mode? 116

2.3.20   Think ahead about editing an image

Unlike many other sorts of documents, technical documentation often needs to be updated multiple times. Each time when there is a new version of the product, there also needs to be a new version of the documentation. These updates are particularly frequent with software.

For this reason, it's very important that you create your images in a way so that you can later edit them freely. Save all images in a format that when you later reopen the file:

- lets you change texts
- lets you edit, move, and delete graphical objects, such as lines, arrows, etc.

Keep the original files

Unless you use an image editor or a version control system that saves backup copies automatically, make a manual backup copy of the initial image. This will enable you to start all over editing at any time.

There are many reasons why you may need this backup copy. For example:

- You've cropped an image. Now you find that it would be better to show a bit more.
- You've downsized an image, reducing the number of pixels. Now you find that the image looks blurry because the resolution is too low. Or you might like to enlarge the image and thus now again need a higher resolution.
- You've applied some image effects or filters that you now want to undo.
- You've converted a color image to grayscale but now find that you also need it in color for your web site.
- You'd like to reuse some part of the image as a stand-alone image. However, this detail is rather small. So you now need the full resolution of the original image to yield sufficient quality.
- You've repeatedly saved the image with lossy compression. Now the quality is too poor.

Keep texts and objects editable

When you save a raster image in a standard format, such as PNG, all layers and objects are flattened. The resulting image consists of nothing but pixels. When you later reopen the image, you image processor only sees these pixels. You can no longer select any objects that you had edited the last time. You can only delete an entire area, leaving a big "hole" in your image.

Images in general

Most image editors have their own, proprietary file formats that are able to store the images in a way so that they remain fully editable. However, you cannot use these proprietary formats directly in most authoring tools and in online documentation. Therefore, you need to create and manage two files:

- the proprietary image file as a backup copy in case you need to edit the image later
- an export of the image into a common file format, such as PNG

Tip:
Some more advanced image editors for screenshots facilitate this process by taking a dual approach: These programs pretend to open and save PNG files that remain fully editable. However, in the background, along with each PNG file, the program automatically saves a second file in its own, proprietary format. So, in this case, you don't need to take care of two files manually, but the program does it for you.

Another good option is SVG. Compared to proprietary formats, this file format has many advantages:

- SVG is not only good for purely vector-based images. You can also use it for raster images, such as photos and screenshots on top of which you want to add text or graphical objects.
- SVG is an open format that many image editors can read and write. So you're free to change the image editor at any time if you need to.
- SVG can be directly used with most authoring tools for technical documentation.
- SVG can be directly displayed in HTML-based online documentation without conversion.
- With SVG, all text and graphical objects that you add on top of the raster images are vector-based. Thus, when you enlarge the image, these elements don't become jagged or blurry (however, the raster image in the background does).
- The SVG format is XML-based. So texts added on top of the raster image, such as callout texts, can the translated with any text editor or translation memory system. In online documentation, these texts can be indexed by the search function.

Related rules

▶ *Think ahead about reusing an image* 123

Think ahead about translating the texts in an image 125

2.3.21 Think ahead about reusing an image

> Technical documentation is often supplied both electronically as HTML-based online-documentation and as a PDF file for printing. In addition, technical documentation is often needed in several variants, such as for a light version of a product and for a professional version.
>
> If you can use an image in multiple of these documentation variants, this saves a lot of time and costs.

Basic ways of making images reusable

- Save your images in an editable format so that you can create modified variants of an image at any time (see *Think ahead about editing an image* 121).
- Consider using a special image editor that supports conditional content. Alternatively, consider putting conditional content onto separate layers within the image. You can then create a specific variant of your image by showing or hiding layers.
- Don't embed the images into your document, but include them by reference (most authoring tools provide a corresponding option when adding an image). You can then use the same image in various places of the same document or even in multiple documents.
- If your authoring tool supports it, in the image references use relative paths, not absolute paths. You can then easily clone your project at any time or move it to a different location without needing to change the paths in all image references.

✘ **No:** C:\projects\documentation\mydocumentation\images\sample.png
✔ **Yes:** \images\sample.png

Features of advanced authoring tools

Consider using advanced image editors and authoring tools that provide special functions for reusing content.

Some image editors that are specialized on technical documentation can, for example:

- use text variables—depending on which document variant you publish, the corresponding texts are filled in
- use conditional content—depending on which document variant you publish, the corresponding elements or texts are shown or hidden

Images in general

- export texts for translation and import them back after translation (see also *Think ahead about translating the texts in an image* 125)

Most authoring systems can:
- reuse snippets that contain images multiple times
- make images conditional so that these images do or do not appear in a particular variant of the document

> **Related rules**
>
> *Think ahead about editing an image* 124
> *Think ahead about translating the texts in an image* 125

2.3.22 Think ahead about translating the texts in an image

Technical documentation often needs to be translated into multiple languages. If your images contain texts, these texts also need translation.

When translating a document into many languages, even small savings in time can add up to significant values.

A key requirement for translating the the texts in an image is that your image files are fully editable (see also *Think ahead about editing an image* 124).

Frequent pitfalls

Translating image texts can be challenging, time-consuming, and thus costly. The typical pitfalls are:

- Translators aren't graphic artists. Normally, they don't have the software needed for editing image files, and they don't have the expertise either.

    Thus, better just send them the plain texts.

- When you send a translator only an extract of the texts of the images, these isolated texts lack their context. This makes translation very difficult and error-prone.

    Thus, better send the original images and documents along with the extracted texts. However, because most translators are paid for the number of words translated, few translators are willing to invest much time into looking at this extra information. Speak about this with your translators, and consider paying them separately for the extra time.

- Texts in other languages are often longer than in the original text (most languages are significantly longer than English). So the texts in the image get truncated, result in awkward line breaks, or cover important parts of the image.

    Thus, review all translated images carefully, and edit them if necessary. This can mean a lot of extra work if you have many images and many languages. Therefore, schedule this step in time.

Allowing for longer texts right from the beginning

When adding text fields to an image, make them large enough (or the typeface small enough) so that there's enough room for translations that are longer.

Purposefully left-align or right-align the text within its text field.

Make sure that growing text does not cover any important areas of the image.

Images in general

✘ No:

Texts have different lengths in other languages.
Texte haben in anderen Sprachen unterschiedliche Längen.
Les textes ont des longueurs différentes dans d'autres langues.
Los textos tienen diferentes longitudes en otros idiomas.
Os textos têm comprimentos diferentes em outros idiomas.
Тексты различной длины на других языках.

✔ Yes:

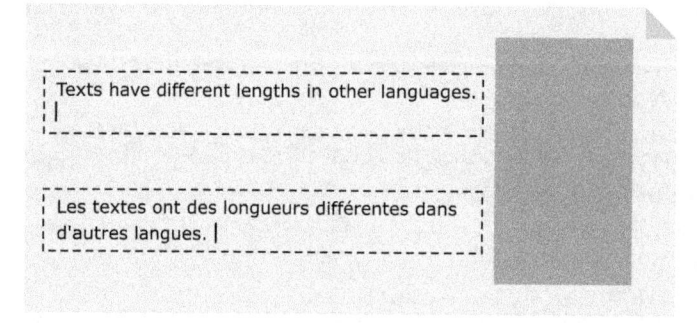

Texts have different lengths in other languages.
|

Les textes ont des longueurs différentes dans d'autres langues. |

Extracting texts for translation efficiently

You don't always need to manually compile a list of all texts that you need to send to your translator. Professional image editors for technical documentation normally have some built-in support for this. The workflow is as follows:

1. You export a text file or XML file that includes all texts and then send this file to your translator.

2. The translator translates the texts within the text file or XML file, using any text editor or translation memory system. Then the translator sends you back the translated file.

3. You make a copy of the image, which will become your translated version.

4. You open the copy in your image editor and import the text file or XML file. This replaces the original texts with the translated ones.

Some image editors already use (proprietary) XML for their source files. Here you don't need the export and import procedure, but translators can edit the

Images in general

text in the source files right away. The same applies if your images are SVG files because internally each SVG file is in fact an XML file.

Minimizing the number of source files

An alternative to working with one image file for each language is to work with multiple layers within the same image. If you have one layer for the texts of each language, you can switch languages simply by hiding and showing layers.

- The advantage of this approach is that in case you make changes to the (language-independent) graphical elements of the image, you don't need to copy and paste these into multiple files but have them all in one place.
- The disadvantage of such "multilingual" images is that you can't send them to multiple translators at the same time.

> **Related rules**
>
> *Include text elements when necessary* 33
> *Be aware of cultural differences* 41
> *Think ahead about editing an image* 124
> *Think ahead about reusing an image* 123
> *Tips for translating screenshots* 214

127

2.3.23 Tips for developing graphics on your own

> Many people think that they don't have the talent and artistic skills for creating attractive graphics.
>
> Fortunately, this isn't so much of a problem with graphics used in user assistance because this type of graphics benefits from being simple. In fact, it's a win-win situation: The simpler the artwork is designed, and thus the easier it is to create, the better it is for communication.

The hard part isn't the act of drawing

The hard part of developing a graphic often isn't the execution. Often, the main challenge is to find the right idea of how to depict the message in a visual form. But it can also be fun and a nice break from writing.

To get creative:

- Go to a place that you like. Listen to your favorite music.
- Take some sheets of paper, a few colored pens, and just start drawing. Experiment. Don't make it beautiful. For now, it's just about the rough idea. You can make it perfect later.
  Working on paper is often more creative than working with an image processor or than drawing on a tablet. It frees your mind from having to think about how to work with the technology. Also, the change of the medium from electronic to Stone Age can boost your creativity.
- Only when the final idea has emerged, sit down at the computer and create the actual image.

Your audience is key

Consider your audience. Use the image types and styles that your audience is used to.

By using the ways of presentation, diagrams, symbols, and so on that your audience are familiar with, you can take advantage of their prior learning. This makes understanding your images easier. In this case, users don't need to learn new conventions but can fully concentrate on the message.

Example: Some audiences are used to particular types of flow charts while other audiences are not. So you would provide a very specific flowchart for one audience while you would provide a basic one for others.

Speaking the visual language of your audience also makes your documents authentic. This motivates users to read your documents very closely because they expect an authentic document to be highly useful. However, if users would get the impression from the graphics that the document is either "too difficult," "just trivial," or "unrealistic," many wouldn't read it at all.

Make users feel positive

Make the viewing of the graphic pleasant:

- Don't frighten users with complexity and information overload. Keep the image as simple and stupid as possible.
- Clearly emphasize what's most important in the image by using bolder lines, darker and more saturated colors, and a central position.
- Prefer soothing colors. Avoid harsh contrasts.
- Use rounded corners rather than sharp-looking ones, which may subconsciously be perceived as being something dangerous.

Consider how often users will consult an image

Even within the same document, the requirements aren't the same for all images.

If users use an image only once, don't expect them to be willing to spend much time with the image. Thus, optimize such an image for speed:

- Design the image to be as simple as possible.
- Embed the image into the document right where it's needed.

If users are likely to use an image frequently, such as a chart for looking up important values, you can expect the users to become more advanced over time in using the image.

- Add more details if these details are needed by advanced users. (If possible, layer the levels of detail. Make the basic and the frequently needed information visually more dominant than the more special information.)
- Make the image easily accessible. Provide enough links to the image. In a printed document it can make sense to put these special types of images into a distinct section of the document, such as into an appendix.

Utilize modern technology and the Internet

When creating the actual graphic, there are a few tricks that can make your job much easier—especially if you aren't a talented graphic artist by nature.

Obtaining an idea for a symbol

To obtain an idea for an icon or symbol, search the Internet for examples (use the image search in a web search engine).

Images in general

Use the example that best matches your design as a starting point for your symbol. However, don't simply copy it. Rather, build upon it—but not by adding things and by making the symbol more complicated but by deleting and simplifying things.

Drawing the shape of an object

If your drawing talent has its limitations, it can be a challenge to draw a neat outline of a certain object. A little technical help can make this job easier.

To obtain a minimalist shape of an object:

1. Take a photo of the object.
2. In an image editor, add the photo on a background layer.
3. Add a new layer, and add some transparency to this layer so that the background image shines through.
4. Use the pen tool or line tool to trace the shape.
5. When the shape has been closed, fill it with color.
6. Remove the transparency of the layer.
7. Delete the background layer with the photo.

Other ways of "drawing" an object are:

- You may apply an image filter that automatically transforms a photo into something that looks like a drawing.
- You may use an automatic tracing function that converts a raster image into a vector image.

> **Related rules**
> *Tips for working with graphic artists and photographers* 131

130

2.3.24 Tips for working with graphic artists and photographers

> Don't expect a graphic artist or a photographer to develop the basic idea of an image for you. This rarely works out well.
> - *You* need to tell them what you want.
> - *You* need to tell them how you want it.
>
> However, you don't need to tell them how to implement it.
>
> The more precisely you can communicate your requirements and ideas, the better. Best prepare a sketch that you can hand over, no matter how simple and awkward this sketch is.

Brief the artist in detail

The graphic artist or photographer can only create a good image if they know enough about the subject and about the specific requirements. Therefore, brief them carefully:

- What's the primary purpose of the image: To inform? To help make a decision? To motivate?
- What exactly must the image communicate?
- Who exactly is the audience?
- What impression do you want to project of your organization and product: modern / conservative / precise / friendly / playful / creative / ...?
- What are your ideas for the image?
- What text may or must be included? What are the exact correct names of the objects that are to be labeled in the image?
- What size will the image have in the document?
- Which fonts are allowed?
- Which colors are allowed?
- How will the image be reproduced? (Online? PDF? Print? If print: Which printing method and color mode—see *Which color mode?* (116))

Make it very clear that aesthetics is secondary

Many graphic designers and photographers have a strong artistic bent and therefore sometimes tend to sacrifice readability and usability to the aesthetics of an image. For example, instead of choosing a large, easily readable font,

Images in general

they choose a small, unobtrusive one. Or they prefer gray text and subtle color variations over clear contrasts that make it easy to read and recognize texts and other parts of the image even under difficult lighting conditions.

So communicate clearly that you need simple, clear images that convey the content as quickly and reliably as possible. This does not mean that the images should not look appealing, but when in doubt, aesthetics are clearly secondary.

Make agreements on copyright

Write a written contract stating who owns the right to use the image (see *Avoid legal pitfalls* 44).

> **ⓘ Important:** Note that payment for the creation of an image does not automatically transfer the image rights to you. For example, even if you pay a photographer to take a photo, this does not automatically mean that you're allowed to publish the photo on the Internet. The artist always remains the creator of the image—no matter whether the work was paid for or not. Unless the artist transfers particular usage rights to you, all rights remain with him or her, not with you!

Review the results carefully

Critically evaluate the content of all images. The graphic artist is the expert for drawing, but you're the expert for the correctness and didactic quality of the images.

Don't judge the images only by their visual appeal. Look at them from a user's perspective:

- Are they easy to understand, based on the prior knowledge of the audience?
- Are they simple? Do they communicate their messages quickly and without the risk of being misinterpreted?
- Are they consistent?
- Are they easy to update? (See *Think ahead about editing an image* 121 .)
- Do they work also in other countries and cultures?
- If they contain any text: Can the texts easily be translated? (See *Think ahead about translating the texts in an image* 125).
- If they contain any text: Are the texts clear? Are there any typos? Do the texts use the same terminology as the body texts of the documents into which the images are to be embedded?
  (See *Tips on writing text within visuals* 37).

Images in general

Related rules

Tips for developing graphics on your own 128

2.4 Images of hardware

Unless your product is mere software, you need to use photos or drawings to show your product and its components.

The following tips will help you decide which types of images to use and how to best create them:

Photo or drawing? 136

Which perspective? 138

Ways of showing the interior of objects 143

Ways of indicating dynamics 146

Prepare a photo session with care 150

Tips for taking photos of technical devices 152

Tips for editing photos of technical devices 164

All general principles for working with images apply as well. Therefore, also see:

Common basics of visualization 19

Images in general 47

2.4.1 Photo or drawing?

When showing your product, you can use either a photo or a drawing. No matter which way of presentation you choose: Be consistent. Standardize your types of images. Always use the same type of image for the same purpose. Don't show the same object alternately as a photo and as a drawing, but stay with the display mode once chosen.

It cannot be generally said whether photos or drawings are more effective. Both have their specific strengths. However, along with the growing popularity of instructional videos on the internet, photos are gaining preference—in particular among younger audiences.

In general:

- Use a photo to show great detail and when your audience may have trouble interpreting a drawing.
- Use a drawing to reduce the amount of detail, or to show details selectively, or to show hidden details.

A drawing must be simple for its strengths to outweigh those of a photo. If you can't make a drawing simple, better use a photo if you can.

Advantages of photos

- Photos show exactly what the product looks like. This minimizes the risk of misunderstandings.
- Photos show parts in their exact positions and perspective.
- Users don't need to perform any mental transfer, which can make interpreting easier.
- Because they look more real, photos better attract attention than abstract drawings.
- Because they look more real, photos are more memorable than drawings.

Advantages of drawings

- Drawings can be simplified to show only what's important. Irrelevant elements can be omitted even if they can't be dismantled in reality. Thus, drawings have particular strengths if they can be simplified in contrast to a photo.
- Drawings can be designed so that parts of an object's interior are visible (see *Ways of showing the interior of objects* 143).
- Drawings can depict even very large objects that don't fit on a photo.

- Drawings can already be created while the product is still under development or doesn't even exist or if the product isn't available for taking photos.
- Drawings can be updated to some degree if the product changes.
- Texts in drawings can be translated, which may be needed if a drawing shows text that's printed onto the product.

Creation costs

Photos are often easier to obtain than drawings and thus cheaper, but this isn't always the case.

- If you have some existing CAD data, you can create high-quality 2D and 3D drawings from this data. However, note that CAD drawings typically are too complex for user assistance and therefore need to be simplified—which involves some extra work.
- If you have some existing photos but prefer drawings, in your graphics software you can try to use an image filter that automatically converts a photo into an image that looks like a line drawing.

Another important consideration can be the question as to how easily you'll be able to obtain additional images of the same kind in the future.

2.4.2 Which perspective?

Show objects as users typically see them. This maximizes the likelihood that your image will be properly understood and that users will perform all actions correctly.

If users might see an object from different perspectives, prefer the perspective that:

- shows the objects in the position of the operation that's described in the text
- is the most common one
- depicts the object as best as possible

If an image shows more than one object, show all objects from the same perspective.

If users can identify an object only by looking at the object from multiple perspectives, include an image from each of these perspectives.

In subsequent images, don't change the perspective unless you tell the user to change the perspective as well. Example: If you tell the user to turn an object, in the next image show the turned object.

Avoid two-dimensional drawings

Unless your audience is very technical, avoid all sorts of two-dimensional drawings of real objects, such as front view, side view, top view, series of orthographic two-dimensional images, and cross-sectional views. These kinds of technical drawings are highly likely to be misinterpreted by many people.

Better, use three-dimensional drawings or photos.

Note:
This does not apply to symbols (see *Use symbols sparingly* 28).

Standard perspectives in drawings

Drawings can use different perspectives to create a 3D effect. When working with a professional illustrator (or when exporting from CAD), you may be asked which perspective to use. So here are the key facts on the most common standard perspectives in a nutshell.

Images of hardware

Note:
The edges of invisible areas are often drawn with a slightly stronger line than other edges (see the examples).

Isometric view

This is the basic parallel perspective. Edges are parallel. Both angles are equally 30 degrees. The lengths are equal in all directions, which typically gives the user a good sense of the dimensions of the object and its components.

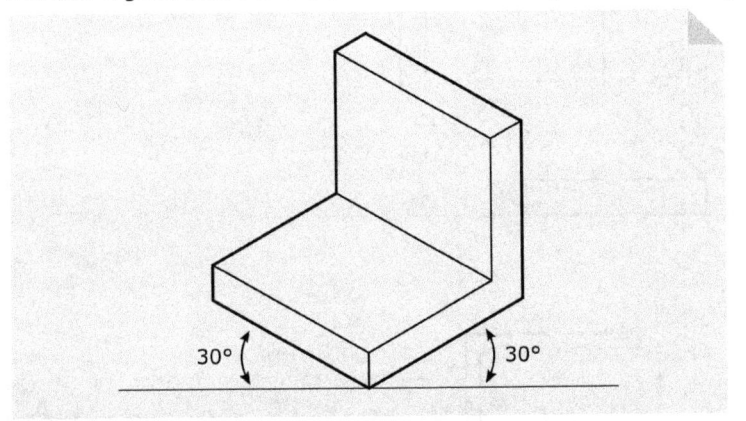

Dimetric view

This variation of the parallel perspective by definition uses two different angles: 7 degrees on the left and 42 degrees on the right. This gives the user a more realistic impression than the isometric view. All lines of the 7-degree angle have half the lengths than those of the 42-degree angle and vertical lines. This saves space on paper or screen.

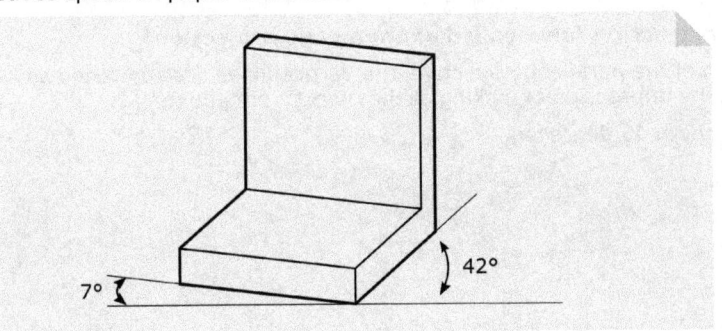

Cavalier projection and cabinet projection

These variations of the parallel perspective emphasize the front of the object. The left angle is 0 degrees; the right angle is 45 degrees. The lines of the 45-

139

Images of hardware

degree angle can either have the same length as the horizontal and vertical lines (cavalier projection), or they can have only half the length to save space (cabinet projection).

Cavalier projection:

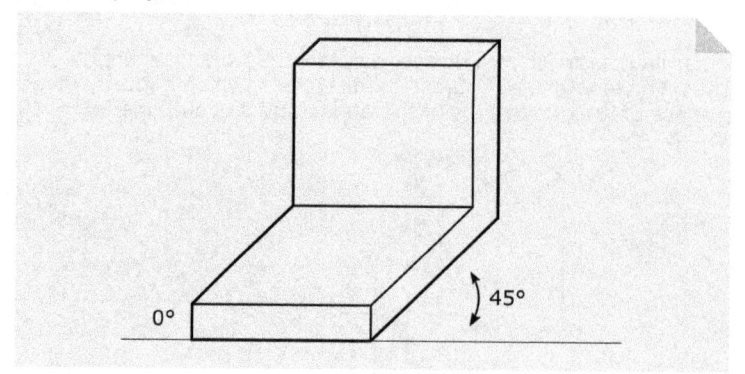

Cabinet projection:

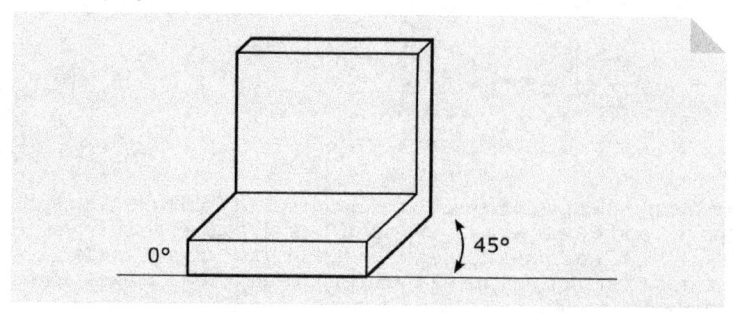

Planometric projection (also called axonometric projection)

In this variation of the parallel perspective, the vertical lines are shortened so that users get the impression of looking at the object from above.

The angles are both 45 degrees.

Images of hardware

Point perspectives

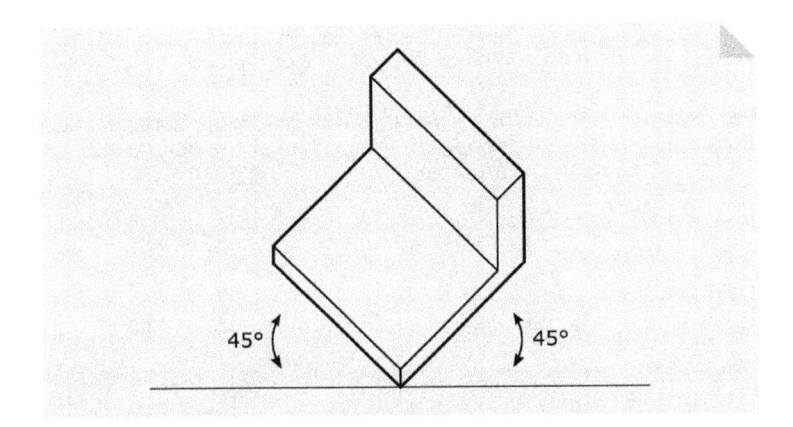

Point perspectives are used to create images that look particularly realistic. The downside of these perspectives is that users cannot clearly determine the dimensions of an object from the lengths of the lines in the image.

A central 1-point perspective with one vanishing point (VP) can be used to create the impression of looking at the object exactly from the front:

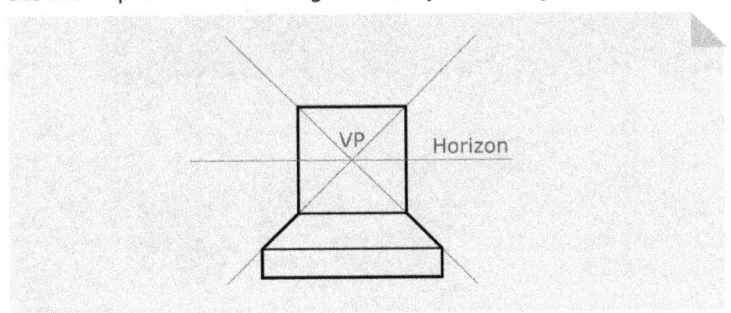

A 2-point perspective with two vanishing points (VP1, VP2) typically looks most natural:

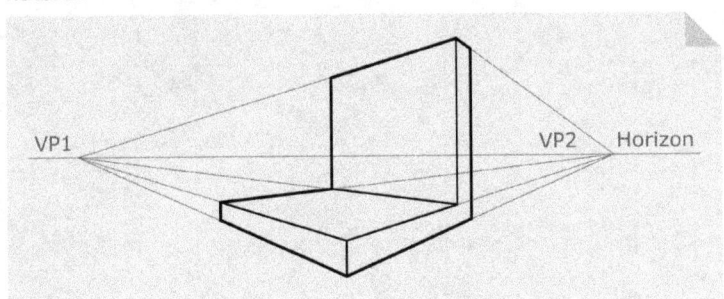

By moving the horizontal line, you can show the object from above or from below.

By adding a third vanishing point in the center above or below the horizon line, you can create an impression of looking far up or far down at the object. In this case, the vertical lines then point towards the third vanishing point

2.4.3 Ways of showing the interior of objects

When explaining how something works or how something is assembled, you sometimes need to show the interior of an object.

Being able to do this is one of the key strengths of drawings compared to photos.

However, because showing the interior makes the drawing more complicated, only show as much of the interior as needed.

Cutaway view (sectional drawing)

A cutaway shows a small section of the interior through a break in the surface of the exterior.

Use this technique if things both in the interior and the exterior of the object are important.

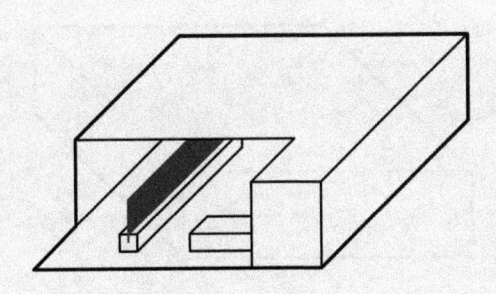

Phantom view (X-ray image)

A phantom view draws the hidden parts in full and only superimposes the outline of the surrounding object.

Use this technique if the interior is more important than the exterior.

Images of hardware

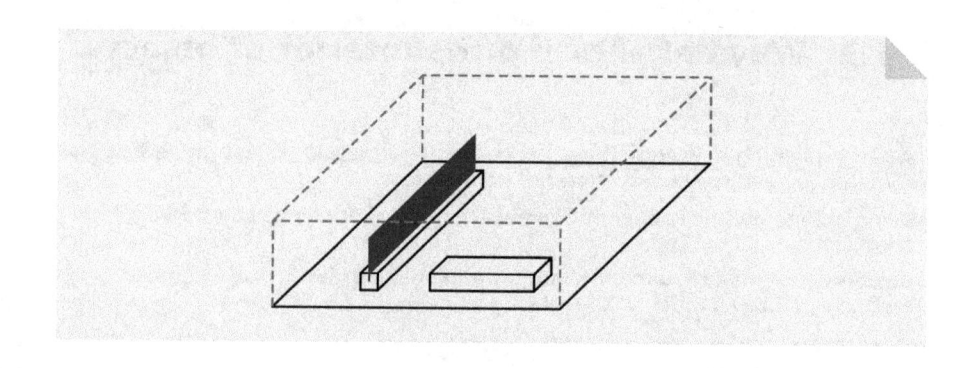

Exploded view

An exploded view shows a device separated into its components.

Use this technique if you need to show all components in detail.

The downsides of this technique are that it needs the most space and that a large image can get quite cluttered and difficult to read for untrained users.

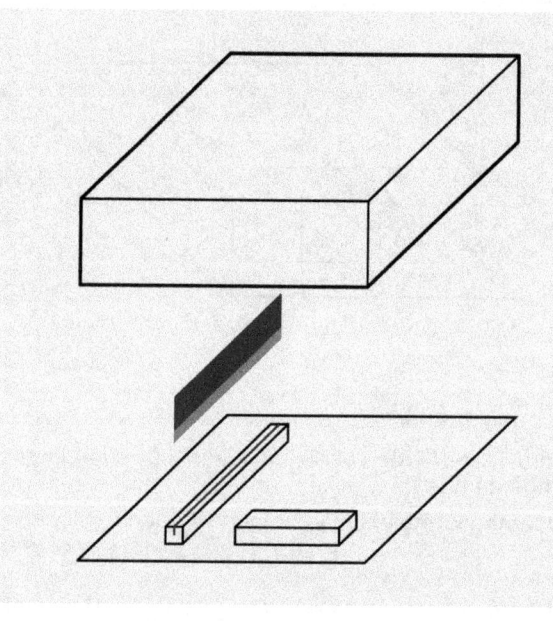

Exploded views are often used to show how to disassemble a device and how to put it back together.

Lines can indicate the direction along which parts need to be assembled.

Branching lines can show alternative assemblies.

Callout lines can point to part names or part numbers (needed for ordering spare parts).

2.4.4 Ways of indicating dynamics

A frequent challenge with images used in technical documentation is the need to indicate some form of movement or change even though the image itself is static.

Most solutions for this use some kind of lines and arrows.

Indicating motion

The typical methods of showing motion originate in the ghost images and motion blur that a fast-moving object creates on the retina of the human eye or on a photo. (Imagine watching a racing car driving by. To the eye, it looks as if a kind of glow or fog follows the car, which are the "ghost images" of the car.)

Ghost images

If the object has a simple outline and surface, you can use actual ghost images. Often, this technique is combined with some sort of arrows.

For simple movements, one or two ghost images can be enough, but you can also add more to indicate an entire motion path.

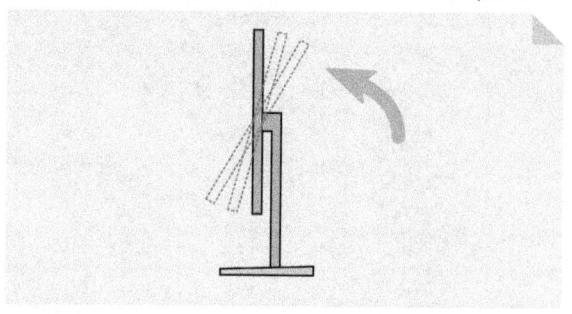

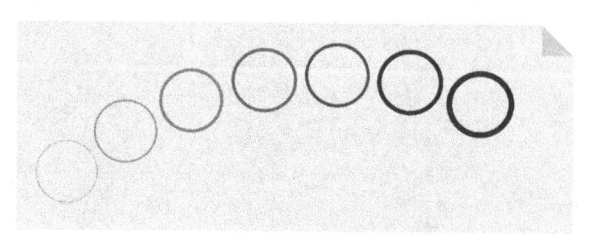

In a photo, the equivalent of ghost images is motion blur. You can use this effect to indicate motion intuitively. The downside, however, is that the moving object can no longer be seen clearly. (To make motion blur stronger, increase exposure time. Also, most advanced image editors let you add motion blur artificially.)

Speed lines

Speed lines are essentially a simplified version of ghost images.

Arrows

Arrows can be used to indicate movement either alone or in combination with other techniques.

Keep the arrows as small and unobtrusive as possible. Don't make them visually more dominant than the subject in the image.

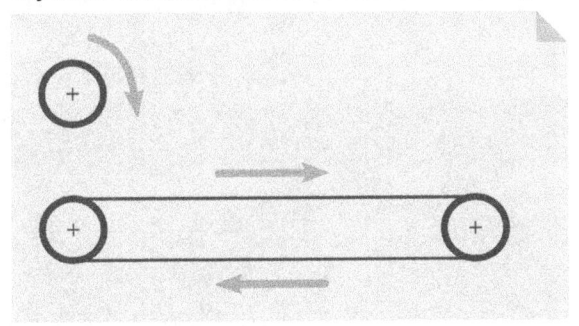

Note:
If you use arrows also for directing the users' attention, visually distinguish the different types of arrows.

Alternative: Using a series of images

An alternative to indicating the movement in one image is to use multiple images, each showing one step in the motion process.

- The advantage is that this approach can show the motion clearly and in detail.
- The disadvantage is that this approach needs more space and may take longer to read.

Images of hardware

If the order isn't clear from the reading order within the document (upper left towards lower right for western languages), add numbers or arrows to indicate the sequence.

Indicating vibration and sound

Shake lines suggest vibration. Essentially, these lines are also ghost images of the vibrating object.

Because vibrations often cause noises and because loudspeakers also vibrate when emitting a sound, shake lines are typically used for visualizing sounds as well.

Example for vibration:

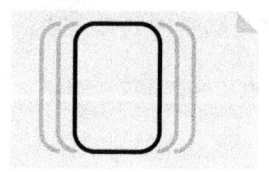

Example for sound:

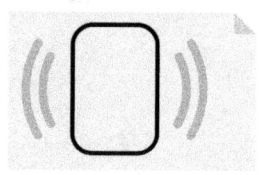

Indicating light and flashing light

Illumination lines can suggest light. These lines are intuitive because they reproduce the reflection effect that a strong point-shaped light also produces on the eye.

To visualize that a light is flashing, you can show the lamp alternately with and without illumination lines.

Images of hardware

Note:
In a photo, you can create quite similar effects with particular image filters that most image processors provide at a click of a button.

Indicating the passage of time

To indicate the passage of time, typically a clock symbol plus text are used.

Duration	Example
Less than one hour	10 min
Less than 12 hours	2h 30 min
More than 12 hours	25 h

149

2.4.5 Prepare a photo session with care

> Don't underestimate the necessity to carefully plan each photo session.
> - Later editing can be much more time-consuming than taking a good photo in the first place.
> - Sometimes, you don't get a second chance to take a missing photo or to retake a poor one because the product is no longer available for a photo session. For example, a custom-built machine may already be on its way to the client.
> - Sometimes, you don't get a second chance to take a missing photo or to retake a poor one because the product has changed its state. For example, once the only machine that is available for taking photos has been assembled you can't retake photos of the assembly.

Use a documentation draft as a "storyboard"

Have a plan as to which images exactly you need. Best have a draft of your document ready before taking the photos. In this document draft, add placeholders for the photos plus comments that specify:

- what exactly to show
- which state the subject must be in (parts attached, positions of switches, etc.)
- which level of detail to show
- which perspective to use

Note:
In case you can't work out a documentation draft beforehand, you'll need to take a lot more photos. Take them from different perspectives in the chronological order of the operation. So you can later write the instruction according to this sequence and decide which steps you can explain with text and which ones need one of the images. This approach isn't ideal but sometimes can't be avoided if the photo session is also your first and only chance for performing or watching the operation.

Helpful preparations

Even though some of the following points may seem trivial, they are often forgotten:

- Make sure that you have free access to the product. If it's a machine in a productive environment, make sure that any standstill of the machine doesn't interfere with the production schedule.
- Make sure that you have the correct version and configuration of the product.
- Make sure that the product and the room are clean. For example, check for oil stains, fingerprints, forgotten packaging material lying around, drilling chips, ….

    Don't forget to recheck for cleanliness repeatedly also during your work—even when in a hurry. Typically, this is particularly important when taking photos of an assembly procedure.
- Make sure that the product is fully set up (unless you're documenting the assembly). Make sure that no parts are missing.
- If you're going to need any samples, demo objects, tools, or instruments: Make sure you have them available and prepared as well.
- Remove any labels or other objects that you don't want to appear on the images.

    Don't forget to also check the background: Are there any irrelevant cables, tools, boxes, pallets, forklifts, bottles, coffee cups, piles of paper, dead plants, pens, jackets, or whatever?
- Prepare the lighting.
- If you don't have much experience: Practice how to use your camera.

> **Related rules**
>
> *Tips for taking photos of technical devices* 152
> *Tips for editing photos of technical devices* 164

Images of hardware

2.4.6 Tips for taking photos of technical devices

Tips about good photography could fill a book alone. However, product photography, as it's needed in technical documentation, is only one small part of photography as a whole. What makes it comparatively simple is that the photos should *not* be impressive works of art. Rather they should be as *simple* as possible.

Essentially, what matters in technical documentation is that:

- the images only show what's necessary
- in the images, all needed details are clearly visible (and preferably only these)

You can produce good results even with only some basic knowledge and affordable equipment.

Main challenges

What can be challenging with photos for technical documentation is:

- Some devices are very small.
- Some devices are very large.
- Many technical devices have polished, glossy surfaces and long straight sides. In the photos, this can result in unwanted reflections and in visible distortions. Reflections also make the correct exposure difficult so that in the images there often tend to be underexposed regions where details aren't clearly visible.

Put the subject on an angle

Typically, products look best if you put them on an angle. When using a straight-on view, forms are often difficult to recognize.

If users can see the subject from 3 sides, this creates a 3-dimensional and thus realistic impression. This helps to clearly identify an object. In addition, it automatically adds some more information: The user learns how the subject looks from *each* side, not just from the front.

Thus, best take photos from a slight angle.

Lighting

Good lighting is a key factor in taking good photos.

If there isn't enough light, this results in so-called *noise* in the image, which makes images look grainy.

The reason for this "noise" is that even if no or only very little light hits the photocell of the camera, the cell emits some electrical charge. This charge is evaluated by the camera as the incidence of light and displayed on the image as colored pixels. In bright areas of the image, these little distortions aren't noticeable. In darker areas of the image, however, the scattered pixels that aren't completely black are visible as little grains.

The longer the exposure time is, the more of these charges occur. Thus, the more noise you have.

Good lighting ensures that the camera can keep the exposure time short, so there's minimum noise in the darker areas of the photo.

For product photography, the ambient lighting typically isn't bright enough. Also, it has a number of disadvantages:

- Sunlight only comes from one direction and thus tends to cast shadows.
- When combined with artificial lighting, the different light colors (so-called light temperatures) result in problems with white balancing.

Also, existing artificial lighting, such as overhead lights or spotlights, can cast nasty shadows.

Bring your own light

For the reasons mentioned above, block out as much outside light as you can and set up your own lighting. Don't mix different light colors, such as existing room lighting and studio lights. This can result in improper white balance and thus in photos that look very unnatural. For the light temperature, typically best choose about 5000K, which pretty much equals daylight.

Use at least two lights that light your subject from different sides. So you get everything visible and avoid shadows.

Use diffuse ("soft") lights—no spotlights. This avoids light reflections on the subject, and it makes cast shadows softer. Best use a set of purpose-built soft studio lights. They are quite affordable.

If you have good lights, you don't need a flash. Because many technical products have plain, glossy surfaces, a flash often results in nasty reflections. Therefore, it's best not to use any flash at all.

Note:
If you can't avoid using a flash: Direct it towards a white surface or an umbrella reflector to obtain diffuse light.

Basic setup

Turn off any overhead lighting and dim the windows (if any). Position two soft lights in front of the subject—one from the left and the other one from the right.

Images of hardware

To avoid reflections, the light should fall onto the subject's surface at a rather flat angle to the camera axis. Approximately 45 degrees typically are best.

Adjust the lights' height so that they are slightly higher than the subject. Any light that doesn't come from above (like sunlight does) looks very unnatural.

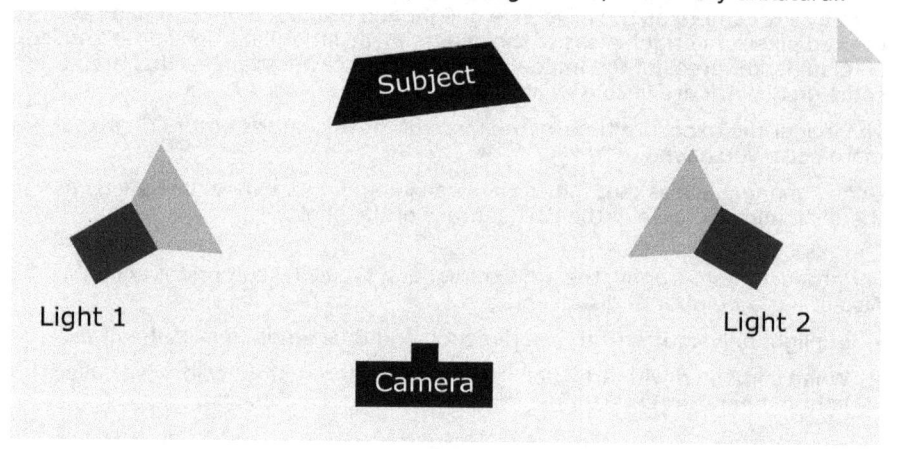

If you want to have the image look particularly 3-dimensional, you can make one of the lights stronger than the other.

To make the light even softer and brighter, feel free to add additional white panels and reflectors around the subject.

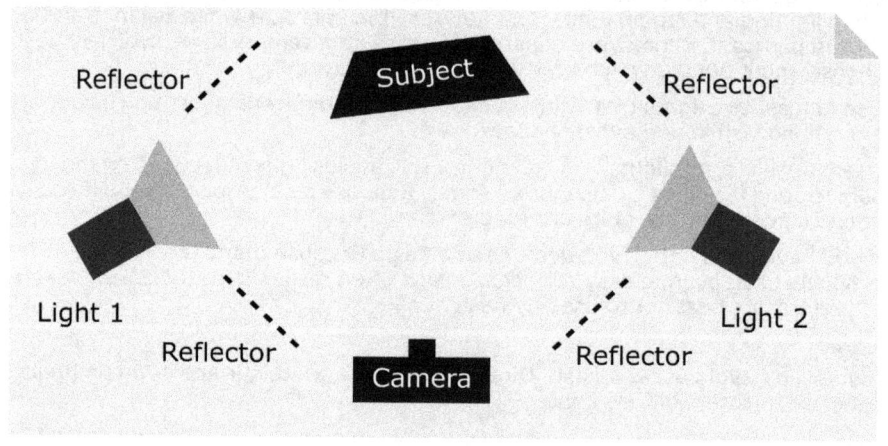

Option: Adding more contrast if needed

Sometimes, you want to highlight a particular detail in an image. In this case, it can help to increase the contrast by using an additional light from the side, which casts small shadows on the surface. You can also use this method to make the structure of a surface more visible, or an engraved lettering or even scratches.

Sometimes, you may like to highlight certain components of a machine in color. You can either do this later in your image editing software, but occasionally it's faster to simply spray paint on the part before taking the photo. Even in the digital age, such classic, pragmatic solutions are worth considering.

Backdrop

For product photos in technical documentation, best use a monochrome backdrop:

- A monochrome backdrop reduces visual "noise." No irrelevant objects or color changes are visible. So users don't need to mentally isolate the relevant objects in the image from the background.
- If you need to select any parts of the image in your image editor (for example, for cropping or for colorizing), this is much easier on a monochrome background.
- You don't need to worry about hiding things that must not be visible in the background.

Ideally, the backdrop should have the form of a curve so that it looks seamless and doesn't produce any shadows.

If the subject that you want to take photos of is small, the easiest way is to purchase some ready-made product photography kit from a photo supply store or some ready-made shooting table or shooting tent (light tent). However, note that it may be difficult (or even impossible) to physically demonstrate any action on the device while the device is within a small shooting tent.

A more flexible solution is to set up a suitable shooting environment yourself:

Best fix a roll of seamless color paper to a backdrop stand or to a wall. Such rolls and stands are available from photo supply stores as well. Instead of paper, you can also use vinyl or polyester backdrops, which are a bit more expensive but also more durable.

If the device that you want to take photos of is small and light enough to be put on a table, fix the backdrop to the wall and to the edge of the table.

Images of hardware

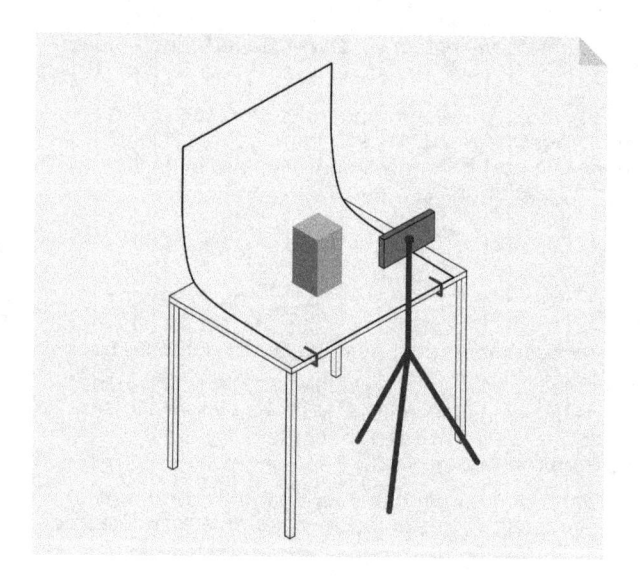

If the device that you want to take photos of is too big or too heavy for a table, fix the backdrop to the floor.

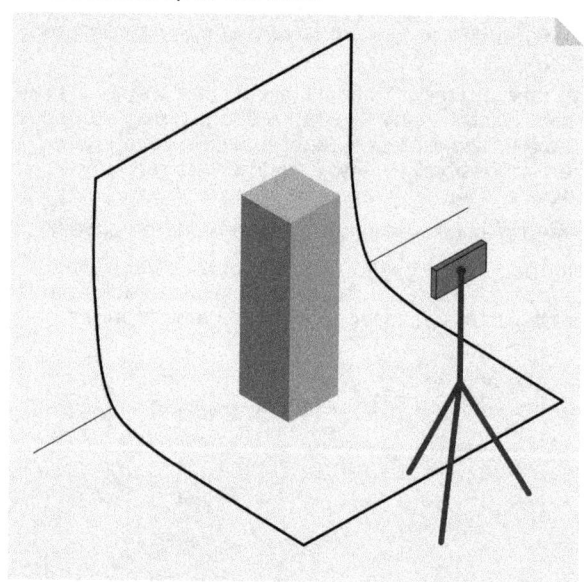

Tip:
You can use the same setups for shooting videos as well.

Setup for shooting photos of large machinery

The larger the machines, the more lights you'll need. Two or three lights often won't be enough here, so make sure to bring enough ones to get everything lit up well.

Eliminating daylight can be another issue. If it's not possible, consider shooting the images at night.

If you're planning to later crop the subject in your image processing software, try to obtain a background that's as monochrome as possible. For example, you can use paper, some cardboard, a tarp, cloth, a wooden wall, or any other largely monochrome material. It doesn't need to be perfect, but it will seriously ease selecting objects in your image processing software with the magic wand or cropping tool.

Understanding how ISO, aperture, and shutter speed work together ("exposure triangle")

A good image is the result of the right blend of ISO, aperture, and shutter speed. On more advanced cameras, you can set these values individually to optimize your image for a particular purpose.

ISO is essentially the camera's sensitivity to light. This is what you should set first.

- At a low ISO, the camera's sensitivity to light is low.
- At a high ISO, the camera's sensitivity to light is high. If the aperture and shutter speed aren't changed, the image gets brighter than at a low ISO. However, the great sensitivity of the sensor also causes more "errors" on the photocell, which result in a grainy look of the photo in dark areas (so-called "noise"). This can become disturbing especially when zooming in to a detail.

Best, keep the ISO as low as possible, and then use the shutter speed and aperture settings to get a good exposure. This will give you rich, smooth colors and low noise.

Aperture is the measure of how open or closed the iris in the lens is. The aperture determines how much light gets through to the sensor, but it also controls the depth of field.

On the camera, the aperture is controlled by the so-called f-stop number.

- A low f-stop number means that the iris is wide open (high aperture). This produces a shallow depth of field: On the image, only the subject is in focus. Everything in the background becomes smooth and blurry.

    This is what you typically need when the important parts of a subject are all within the same distance from the camera and you want the users to focus on these parts and ignore the background. It can also help to hide things that you do not want to be visible on your image—either because you want

Images of hardware

your image to be as simple as possible, or because what's in the back of your office or production hall isn't particularly presentable.

- A high f-stop number means that the iris is quite closed (low aperture). This produces a wide depth of field: On the image, everything both in the foreground *and* in the background is in focus.

  This is what you typically need when there are important parts of a device or machine both in the foreground and in the background.

Note:
With smartphone cameras, you typically can't change the aperture. Normally, smartphone cameras by default give you a very wide depth of field. More advanced cameras additionally have a special function and lens that enable you to take close-ups also with a blurred background ("Bokeh effect").

The **shutter speed** determines how long the light is allowed to hit the sensor.

- High shutter speed means that the light hits the sensor only for a very short time.

  High shutter speed freezes the action. It minimizes the risk of blurring an image because your hands are shaking or because the subject is moving (motion blur). To have the image properly exposed, you need:
  - good lighting
  - a high ISO (that is a high sensitivity to light)
  - a low f-stop-number (that is a wide-open iris of the lens)

- Low shutter speed means that the light hits the sensor for a longer time.

  This may be needed under poor lighting conditions but results in motion blur if the subject is moving. Therefore, you normally only use a low shutter speed:
  - if you're working with a tripod
  - if the subject doesn't move or vibrate

  However, at times you may also like to have some motion blur on purpose to indicate movement with a still image. In this case, low shutter speed is crucial.

Automatic settings

Typically, you don't need to set all values of ISO, aperture, and shutter speed manually. Most cameras let you set one or two values and automatically calculate and set the other. For example, if you want to isolate an object on the image in front of a blurry background, you can just set the ISO and a low f-stop number, and then the camera will do the rest for you. Or if you want to depict the motion of an object, you can set the shutter speed to a low value, and let the camera automatically choose the aperture and the ISO as needed to produce an image without too much digital noise.

Option: Using a telephoto lens

If you need a *very shallow depth of field*, one technique is to use a telephoto lens and to move the camera far away from the subject. However, this only works if you have enough space—which is often not the case in small rooms or with large machinery in a production hall.

Option: Focus stacking

If you need a *very wide depth of field*, the technique of *focus stacking* can give you a perfectly sharp image both in the front and in the back.

The principle of focus stacking is to take multiple images of the subject at varying focus settings without moving the camera. Later the images are merged in some image editing software that supports this technique. To make it work, the camera must not be moved between the shots and the lighting must also remain exactly the same.

Option: Improving contrast and reducing noise with HDR multishot images

High dynamic range (HDR) images are made by taking several photos (typically three) of the same scene, each at different shutter speeds: One bright photo, one medium one, and one dark one. Afterwards, a software algorithm combines all the photos into a single one, which brings more detail and contrast to both the shadows and the highlights.

Avoid using a wide-angle lens

Don't use a wide-angle lens if you can avoid it.

A wide-angle lens can help you to get more on your image when you can't step back far enough from the subject. However, the downside is that there can be lots of distortion at the edge of the frame.

With objects that have many straight lines and edges—as is the case with most technical devices and machines—this can look strange.

Beware of the keystone effect ("converging verticals")

When taking photos of tall objects that have many straight lines, such as technical devices and machines, these objects can look unnatural on a 2-dimensional photo. This is due to the so-called *keystone effect* or *keystone distortion*. Lines that in reality run in parallel then don't run in parallel in the image.

Images of hardware

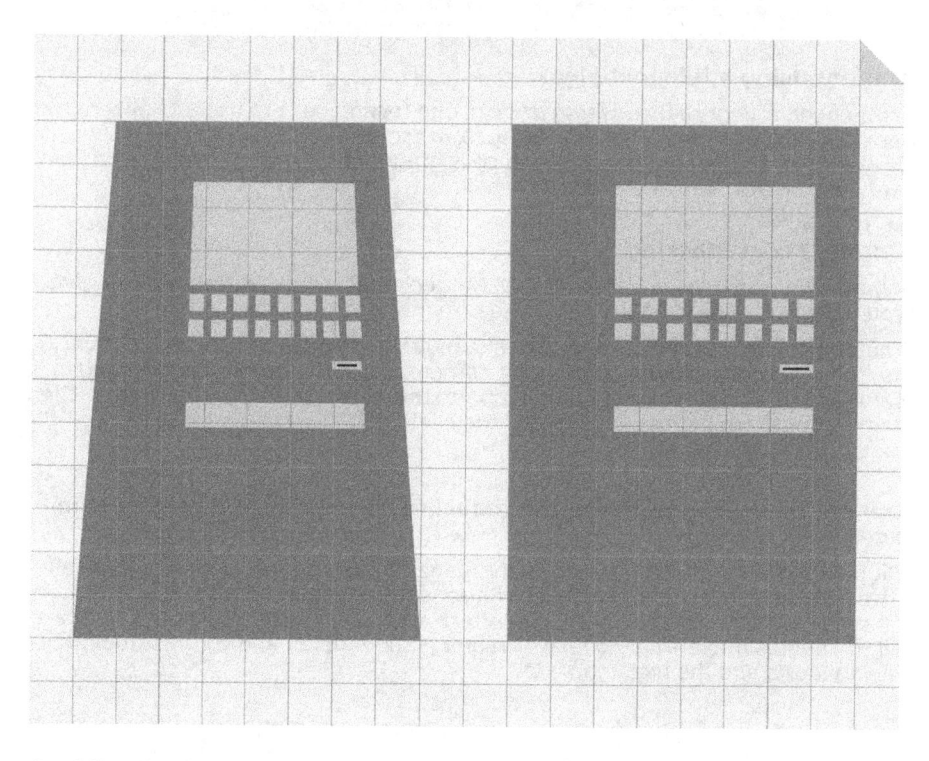

Avoiding keystone effects

To avoid or at least to minimize keystoning:

- Make sure that the sensor of the camera is as parallel to the subject as possible. Don't stand much below or much above the subject.
- If you can't get the camera higher or lower, move further away from the subject if possible. This can be very effective. If you have a good camera and use a tripod, you can usually step back pretty far and still achieve sufficient image quality. If you have one, you can use a telephoto lens to retain the full resolution despite the greater distance.

Tip:
With most cameras, you can show a grid on the screen. Comparing the grid lines to the subject's main lines or edges can help you determine the current degree of keystoning.

Removing keystone effects

Often, it's not possible to move the camera to the optimum position. For example, in a production hall there may just not be enough space to move the camera back as far or as high as you need.

In a case like this, you can later edit the image in your image processing software. More advanced image editors provide some special functions for this task. The downside is that it involves some extra work. Also, the procedure will cut off the edges of the image to some extent.

The worse keystoning is on the original image, the more difficult electronic correction becomes and the bigger is the part of the image that you lose. Therefore:

- Always look out for the keystone effect and take your photos as well as possible right from the beginning.
- When taking your photos, always leave some space around the subject so that when editing the image there's some material that can be cut off without losing information.

Note:
There are also some special perspective-control lenses available. However, they are costly and require special cameras and some particular expertise. For technical documentation, using a standard camera in combination with some good image processing software is usually sufficient.

Pay attention to authenticity

Especially when showing people (or even only their hands performing an action), don't ruin the credibility of your images. If a person in an image looks or acts like a photo model, for technical documentation the image will be worthless or even ridiculous.

- Pay attention that the shown person is wearing the necessary protective gear if needed.
- Make sure that the person doesn't violate other safety regulations, for example by wearing a wristwatch, jewelry, or loose clothing when operating a machine.
- Take care that the person in the image holds or manipulates the object in the way that the object needs to be handled in real-life situations.

Tip:
This doesn't mean that a presenter shouldn't be cleanly dressed. For example, it can be a good idea for the actor to wear a nice T-shirt, overalls, or jacket in your company color and with your company or product logo—provided it looks authentic.

Pay attention to context

When taking a photo of some detail, don't forget that the final photo might need to show a bit context so that users can easily locate the object within the product.

Images of hardware

It's therefore usually better to choose a slightly larger section rather than a too small one. If you work with a sufficiently high image resolution, you can still crop the image later—but you wouldn't be able to add anything missing later.

Things that can establish some context in an image are:

- some known objects nearby
- some characteristic, easily recognizable objects nearby (for example, objects that have a distinct shape or color)
- some additional objects that the users already know so that they can assess the size of the subject (for example, a tool, a wall outlet, a hand, ...)

Use a tripod or other stand for the camera

If you set your camera on a tripod, this ensures a fixed position and avoids camera shake.

- It always gives you the same position. So you can take your time to gradually optimize both the camera and the subject.
- If you need to take several shots that show different steps of a procedure, you maintain a coherent perspective. This eventually makes the instructions that use your images consistent and easy to follow.
- The tripod reduces the risk of blurring the image. Your hands are not as steady as you might think. Even if you don't see the hand shake right away —when you need to zoom in on a detail in the image, you will.

Consider using a gray card

The human eye always sees a white object as white, no matter which light source illuminates it. A camera however, captures the actual light color. Depending on the light source, this color is somewhat different from pure white. Without correction, the image would have a tint.

The automatic white balance of a camera corrects this effect. But it doesn't always work correctly—especially if there are different light sources at the same time. In this case, the white balance needs to be adjusted manually.

A so-called gray card can help with adjusting the white balance. The gray card provides a spectrally neutral surface that reflects equal amounts of red, blue, and green under all light conditions.

To use a gray card for metering before taking an image:

1. Set up your lighting.
2. Place the gray card near the subject and adjust the camera's white balance settings until on the images the gray card has no more tints and looks absolutely plain gray. (This task can be quite time-consuming and imprecise, why many photographers prefer to use the gray card only for color correction in post-production.)

Images of hardware

3. Take your photos. As long as the lighting doesn't change, they now have the correct white balance. In case the lighting changes, you need to repeat the setup of the white balance.

To use a gray card for color correction in post-production:

1. Set up your lighting.
2. Make sure that your white balance on the camera is set to match the dominating lighting type. If you aren't sure which option to use, set it to "auto white balance."
3. In the first image of a series, place the gray card near the subject and make sure that the card is well lit.
4. Take all other images of the series without the gray card. Important: In case the lighting changes for whatever reason in whatever way, you'll need to add another image with the gray card to use for the subsequent images.
5. In post-production, first open the photo with the gray card in your image editor.
6. Adapt the white balance settings of the image until the gray card in the image looks plain gray and has no more tints.
7. Copy the white-balance settings and apply them also to the other images of the series.

> **Related rules**
> *Prepare a photo session with care* 150
> *Tips for editing photos of technical devices* 164

2.4.7 Tips for editing photos of technical devices

With the power of today's image processing software, you don't need to be a graphic artist to accomplish basic photo editing tasks. However, you may find the set of features that typical image processing software provides quite overwhelming.

Of the many features, essentially you'll need only a few basic ones. Try to keep photo editing to a minimum. Good photos are made with the camera, not with image processing software.

Before investing a lot of time into trying to improve a poor image, rather consider retaking the image.

Correcting colors

If an image was poorly exposed, you might need to correct white balance, brightness, or contrast.

Be very careful with changing colors. Don't solely rely on your impression on one screen. On another computer screen, or when printed on paper, colors may look very different.

If an image is going to be printed in black and white, it can make sense to convert color images manually into grayscale rather than having the printing process convert them automatically. With automatic conversion, the problem is that sometimes two completely different colors can result in the same shade of gray and thus significant distinguishing information gets lost. This is something that you can detect and optimize only manually via channel mixing.

Tip:
In your image editor, use the explicit function for converting color to grayscale rather than just removing color saturation. The explicit function normally uses an algorithm that weights the colors according to their perceived brightness. The results then look crisper due to the better contrasts.

Resizing the image

Be careful with resizing images.

- If you scale an image down, the image irrevocably loses pixels.
- If you scale an image up, the image's existing pixels essentially become larger. No matter how sophisticated the enlargement algorithm is: It will never provide a perfectly crisp image.

Take care to lock the aspect ratio so that the image doesn't get distorted.

Trimming the image

If the margins of an image show irrelevant objects, you should trim the image to cut off the unnecessary information.

However, make sure that the trimmed image still shows enough surrounding objects (context) so that it's still clear what the trimmed image shows and it remains obvious where the visible parts are located.

Also, be aware of the fact that trimming always reduces the number of pixels in the image. Thus, to keep the resolution (DPI) constant, you need to make the image correspondingly smaller in your document (see *Which resolution?* 111). Else, it may look pixelated or blurry.

Removing objects

Often, you can't remove irrelevant components of a device physically before taking a photo, or you can't get rid of them by trimming the edges of the image. However, you can remove unwanted things electronically. Users can then better focus on what's important.

Tip:
Use a monochrome background when taking the photo. This can make selecting objects much easier. In your image processor, get familiar with more advanced selection features, such as lasso, magic wand, and magic eraser.

Highlighting objects

If you can't crop an object completely but need to show it in context, you can make it visually stand out from the surrounding objects:

- You can make the object visually stronger by increasing its color saturation.
- You can make the surrounding objects visually weaker by decreasing their color saturation or contrast, or by blurring them to some extent.

Small touch-ups

By using the various drawing tools of your image processing software, you may like to:

- remove dirt or scratches from surfaces
- remove confidential data, such as a serial number
- soften reflections

Images of hardware

Transformations

If an image isn't completely horizontal, you can slightly rotate it.

If you couldn't avoid any keystone effects when taking a photo, more advanced image editors provide functions to remove these effects electronically.

Note:
These transformations will trim your image a bit at its sides.

Related rules

▶ *Prepare a photo session with care* 150
Tips for taking photos of technical devices 152

2.5 Images of software

Screenshots that show the documented software are the most frequently used form of visualization in software documentation.

There are a few things that make screenshots special:
- Screenshots are easy to take. No camera and no special equipment are needed. Also, you don't need to draw a screenshot manually. All you must do to capture a screen is to press a button.
- Screenshots contain text. Other than with most photos of hardware, you need a different version of the screenshot for each language into which you're going to translate your documentation.
- Software is made for processing data. Thus, most software user interfaces also show data—so will your screenshots. It can be much more work to populate the user interface with meaningful sample data than to take the screenshot itself.

  For translations, the sample data may also need to be translated.
- Other than most hardware products, software typically gets many updates during its life cycle. So you'll also need to update many of your screenshots multiple times.

These characteristics result in the following particular best practices for working with screenshots:

How many screenshots to show? 169
Use a real screenshot or an illustration? 174
What to show in a screenshot? 177
When to include the mouse pointer? 183
Don't maximize the windows to capture 186
Get rid of dead space 189
Avoid confusion with the real UI 193
Use standard settings 194
With web applications, don't show the browser 197
Show meaningful data 199
Hide private data 202
If it saves you time, fake a screenshot 205
Optimize each screenshot for its particular purpose 207
Tips for taking screenshots 209
Tips for translating screenshots 214

Images of software

All general principles for working with images apply as well. Therefore, also see:

Common basics of visualization 19

Images in general 47

2.5.1 How many screenshots to show?

For several reasons, a lot of authors tend to add too many screenshots to their documents.

- The screenshots look nice and colorful.
- Taking and adding a screenshot typically doesn't take much time. Mostly, it takes less time than adding an equivalent amount of text. Screenshots can fill a large number of pages quickly, which makes the author happy.
- Somebody once said, "a picture is worth a thousand words." Therefore, many authors think that the more images they include, the better the documentation will be.

Don't fall for the temptations to use screenshots in abundance!

- Only add those screenshots to your documents that add some real value. Simply put, all other screenshots are nothing but useless clutter.
- Be aware of the fact that adding screenshots can be done quickly, but that keeping them up to date, translating them into many languages, and even keeping them up to date in all of these languages can be a time-consuming and costly challenge.

   The total number of screenshots = number of screenshots × number of languages × number of updates.

   Always keep in mind that there are two(!) multiplication signs in this formula.

Advantages of screenshots

- Screenshots can help to better visually structure your text and make the final document visually more attractive. A visually attractive document improves the users' motivation to read the document.
- Documents with lots of text and no or just a few images look intimidating to many users. Screenshots provide some visual relief on pages full of text.
- Screenshots can work as landmarks, which give some orientation to users within the document. Also, screenshots can facilitate random access entry into the document: Users can skim the document for an image that shows a specific window or object.
- Screenshots can give users confidence that they are at the right place doing the right thing.
- Screenshots make it possible for users to understand the descriptions in the documentation even when they aren't sitting in front of the product.

Images of software

All of these points are especially important for beginners but less important for advanced users.

Disadvantages of screenshots

- It can take a lot of time to take a screenshot that contains meaningful sample data. (Taking the actual screenshot, however, is easy.)
- It takes time to update the screenshots. (Even a small change in the user interface might result in having to update a large number of screenshots.)
- It takes time to translate the screenshots or to retake them in other languages.
- It takes time to update also the translated versions in the future.
- Screenshots make your documents longer, which has several disadvantages:
    - Printed documents have more pages, which increases printing and shipping costs.
    - Online documents need more storage space and bandwidth for transmission.
    - Users need to turn more pages or to scroll more often.
    - Documents that have many pages can look intimidating and may create the impression that your product is difficult to use.

Typical use cases for adding a screenshot

There are a number of typical scenarios when adding a screenshot makes sense:

Screenshots can help develop a mental model

Screenshots can contribute to the development of the users' mental model of your product by:

- acquainting users with the main windows
- explaining the spatial layout of the windows
- developing a sense of the logical flow within the program

Screenshots can locate and identify controls mentioned in the text

Screenshots can show the user what a mentioned control looks like and where to find it.

Screenshots can demonstrate how to perform an action

Screenshots can show users how to perform a particular task. Because with software most actions are physically simple and standardized, however, this application is quite rare for screenshots.

Tip:
In online documentation, an animated image or a short video can often do a better job for demonstrating an action than a still image can.

Screenshots can help to determine the state of the system and to verify results

Screenshots let users visually compare a described situation with their own situation.

- *Before* performing a described action, users can use a screenshot to make sure that:
  - their own situation matches the described scenario
  - all prerequisites have been met
- *After* performing the described action, users can verify whether their action was successful. The positive feedback conveys a feeling of confidence and motivation.

Screenshots can help to sell your product

Normally, images shouldn't be added to user documentation for marketing purposes. But sometimes, documentation can take a dual role and also works as a marketing tool: Trial versions are an important sales channel for many software products. So the documentation that ships with the trial version to some extent can indeed also be sales material.

If you do add an image for marketing, do it purposefully. Screenshots can help to sell your product by:

- informing users about your product's look and feel
- drawing special attention to a certain function
- communicating a particular benefit

Conclusions for the frequency of screenshots

Resulting from the pros and cons of screenshots, and following from their typical use cases:

- Add as many screenshots as necessary to your documents but as few as possible.
- Don't attempt to have screenshots everywhere. It's perfectly OK to have topics that don't have any screenshot at all.

- Don't attempt to add the same number of screenshots everywhere. Not all documents, and not all parts within the same document, need the same density of screenshots.
  - Topics that provide reference information typically need few or no screenshots.
  - Topics that explain concepts typically need a medium density of screenshots.
  - Topics that provide task-based information typically need the highest density of screenshots.
  - A printed user manual often requires more screenshots than online documentation. The reason is that users of online documentation are typically sitting right in front of the software anyway. This is not necessarily the case with a printed manual.
  - Not all users need the same density of screenshots. Screenshots that may be helpful to beginners can be annoying and even hindering for advanced users. Thus, a getting started guide, for example, needs more screenshots than a reference manual.
  - If the user interface is complex, you need more screenshots than with a simple one.
- Also, consider what each screenshot costs you in terms of time and money.
  - If you can easily populate your screenshots with meaningful sample data, you may afford to add more screenshots than with a very complex setup.
  - If your document won't need to be updated very often, you may afford to add more screenshots than with a document that undergoes frequent changes. Be particularly careful with new products! New software typically changes quite often and quite radically during the first months (sometimes years) of its existence.
  - If your document won't need to be translated into many other languages, you may afford to add more screenshots than with a document that needs each image in multiple language versions.

You don't necessarily need to show each step

Some authors include a screenshot into their documentation for each step of a procedure. In most cases, this is *not* necessary.

Look at your audience:

- If you think that a substantial proportion of users will have trouble performing the action, *do* include the screenshot. (This is most likely if the users are inexperienced or if your product differs from what the users are used to.)
- If you think that a substantial proportion of users will feel uncertain, *do* include the screenshot.
- Else, *don't* include the screenshot.

You don't necessarily need to show each result

Like with steps, some authors also include a screenshot for each result of an action. In most cases, this is also *not* necessary—neither for users nor for reasons of consistency.

If you can clearly describe a result with a single short sentence, this is usually better because it takes less space and is faster to read. Use an image only if without the image you need a longer description or if without the image your description may be ambiguous.

But also here, in the end it depends on the audience.

2.5.2 Use a real screenshot or an illustration?

In software documentation, it's most common to use real screenshots. However, there are also cases where it's better to use an illustration of a software user interface. An illustration may look very similar to reality, or it may be abstract and simplified.

Using an illustration can be a good solution:

- if you cannot take a real screenshot (examples: the software does not yet exist; the software doesn't work properly; you don't have access to the software; the software runs on a device that doesn't allow you to capture the screen)
- if you don't need to show the screen in detail but only some characteristic elements

Using illustrations can also simplify the translation process (see *Tips for translating screenshots* 214).

Advantages of real screenshots

- Avoids confusion. The image looks exactly like the product.
- Avoids errors.
- Can be taken quickly just with a click of a button.
- Creating the image doesn't require any graphical expertise.

Advantages of illustrations

- Can simplify and abstract what they show.
- Can be created independently even if the software or particular functions are not available.
- Can be translated.
- Can be edited.
- If the image is simplified and doesn't attempt to look exactly like the real software, it won't get outdated as soon as the design of the software or the look of the operating system change. Also, it's completely platform-independent.
- If the image is vector-based, it can be scaled to any size without looking blurry or pixelated.
- If the image is included in SVG format into online documentation, the texts within the image are searchable.

Simplified user interfaces (SUI)

Some screen capture tools can automatically or semi-automatically convert a real screenshot into a simplified one. Here, only the important elements within the image still look real. The rest consists of simple shapes that only roughly suggest what's there.

Tip:
You can achieve the same effect also manually quite easily: In an image editor, add some filled rectangles or other shapes on top of all elements that you wish to simplify. For the colors of the shapes, best choose some colors from the original image. This usually produces a harmonious impression that's as close to reality as possible.

The next image shows an original screenshot:

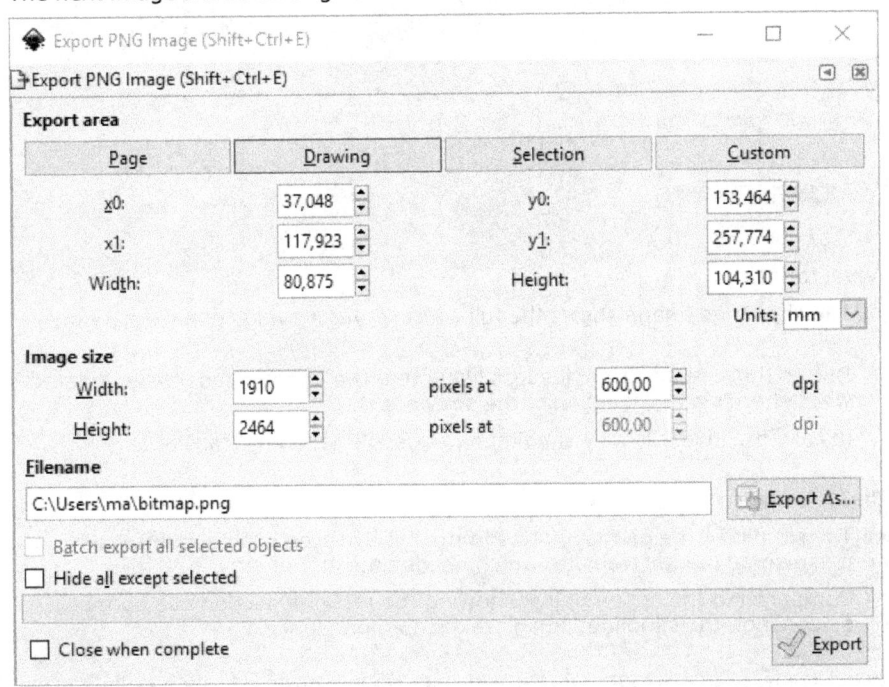

When the original screenshot is converted into a SUI image, it may look like this:

175

Images of software

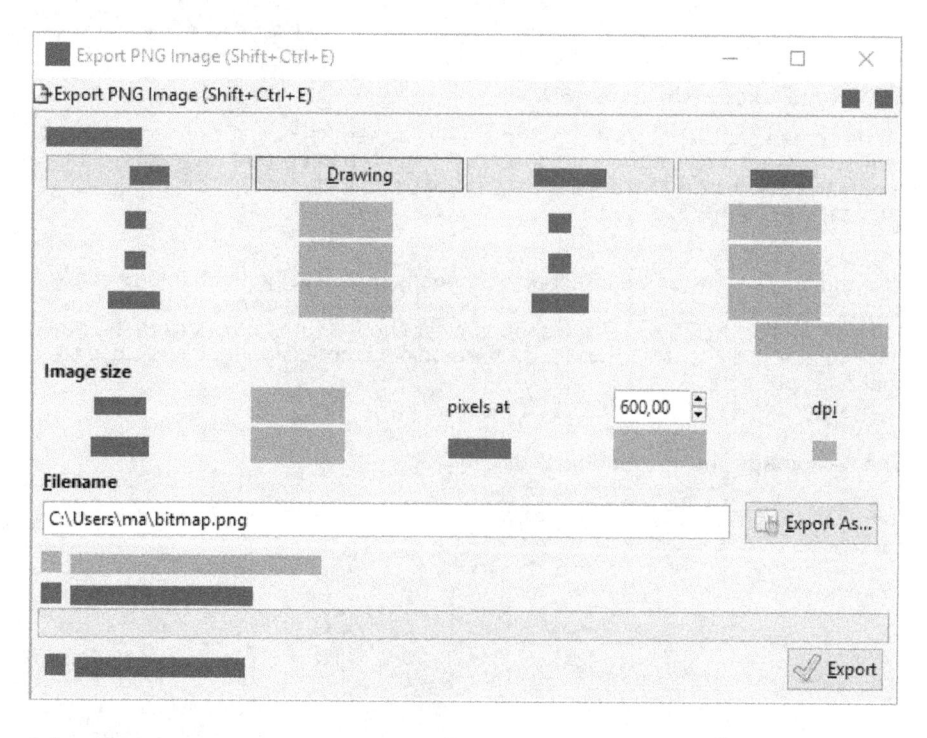

Advantages:

- The simplified image shows the full context, yet it avoids unnecessary detail.
- Due to the simplification, it's less likely that the image needs to be updated when there's a new version of the software.
- In case the image is fully simplified, it doesn't need any translation.

Disadvantages:

- The simplified image may not be immediately recognizable. Users must make some mental transfer, which needs time and energy.
- Compared to the option of just showing the relevant section in a normal screenshot, the simplified image takes up more space.

Thus, SUI images are generally good for explaining concepts and workflows and for developing a mental model of the software. However, they can be poor for describing procedures, where it's often important to show something in detail exactly as it looks on screen.

2.5.3 What to show in a screenshot?

> Should you show the whole screen, a complete window, part of a window, or just a small area or even only a single control?
>
> As with all images, only show what's important for your topic. Crop out all extraneous information.
>
> However, also provide enough context: Make sure that users can easily identify which part of the user interface the image shows, so they know where to act within the product.

Don't show the entire screen unless it's necessary to do so

If you're referring to a particular window as a whole, typically only show this window but not the entire screen with all overlapping windows.

If you're referring to just a part of a window, and if the rest of this window isn't important, consider showing only the part of the window that's relevant.

The advantages of showing only the relevant part of a window are:

- Users can easily focus on what's important and don't need to mentally hide the irrelevant parts of the image.
- The image needs less space on paper or on screen.
- When the software changes with an update, it's less likely that the changes affect the image.

The disadvantages of showing only a part of a window are:

- It's more difficult for users to find the relevant section in the software.
- It's more difficult for users to interpret the image in context.

Provide enough context

When not showing the whole screen or window but only a small section, users may lack orientation.

Therefore, in a cropped screenshot always show enough surrounding elements that make it easy for users:

- to understand which window or panel the shown area is part of
- to spot the shown area on their own screens

Images of software

This is particularly important when describing procedures that users are supposed to follow, and when showing results that users are supposed to compare to their own screens.

Things that can provide the necessary orientation are:

- window titles
- window borders next to the shown area
- overlapping window titles to indicate *how* the window has been opened
- a background window to indicate *where* the window has been opened
- any visually dominant and characteristic elements next to the shown area

For example, including a part of the toolbar and the right window border can clearly indicate that the shown area must be near the upper right corner of the application window. Similarly, including some part of the status bar can clearly indicate that the shown area is at the bottom of the window, etc.

The following example image has no context at all. It only shows how the icon looks but gives no clue whatsoever as to where the icon is located.

The next image includes a bit more context. It shows that the icon belongs to the "Character" group.

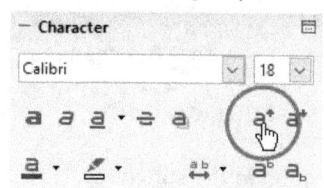

Even more context can be added by showing the window's border at the right.

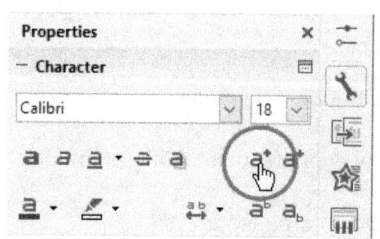

So you must look for the icon on the very right side of the application window. Also, you can now see that the "Character" group is located on the "Properties" panel.

178

Finally, if you also include the upper window border and the window controls, it becomes apparent that the "Properties" panel is located at the upper right of the application window.

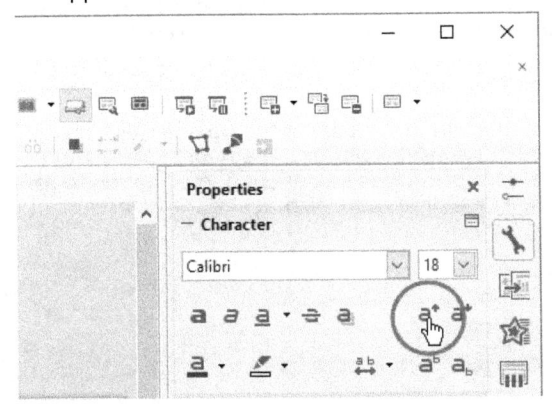

How much context you actually need to show in a particular case depends on the document and its users:

- Beginners need more context than advanced users.

  Thus, tutorials and getting started guides typically contain many full-size images, showing the whole application.

- Advanced users already know the user interface well and therefore only need very few visual cues.

  For this reason, reference manuals typically contain only small, cropped images—if they contain any images at all.

Make sure that it's visually obvious in case an image has been cropped

If you only show a part of a window, make sure that there are enough visual cues that the image has been cut.

Good indicators are:

- cut-through window borders
- cut-through panels
- cut-through lines
- cut-through controls
- fading effects
- torn-edge effects

179

Images of software

Don't confuse users by removing elements from the shown section

Good documentation creates a sense of trust. Trust, however, can be destroyed if the documentation differs from what users see on their own screens.

For this reason, do not delete anything within the shown area (except for dead space—see *Get rid of dead space* 189).

If there are single irrelevant elements within the part of the screen that you're showing, rather than cutting them out consider blurring them, or replace them with an abstract shape or outline. This preserves just enough information to recognize these elements but at the same time reduces much of the attention that these irrelevant elements attract.

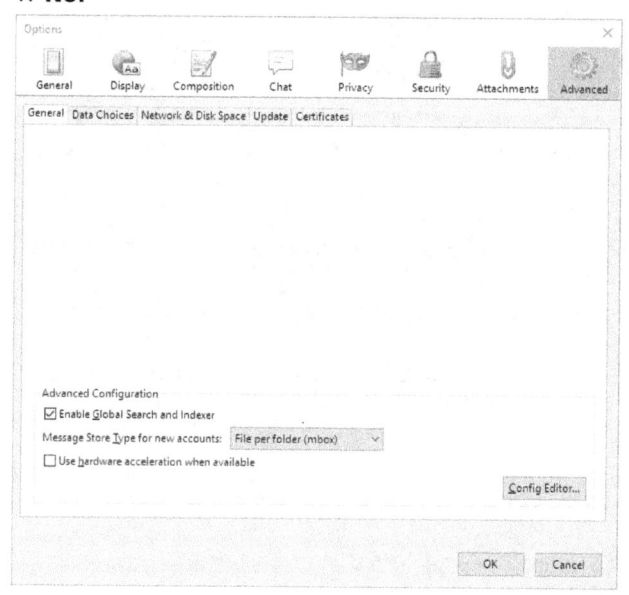

✖ **No:**

Images of software

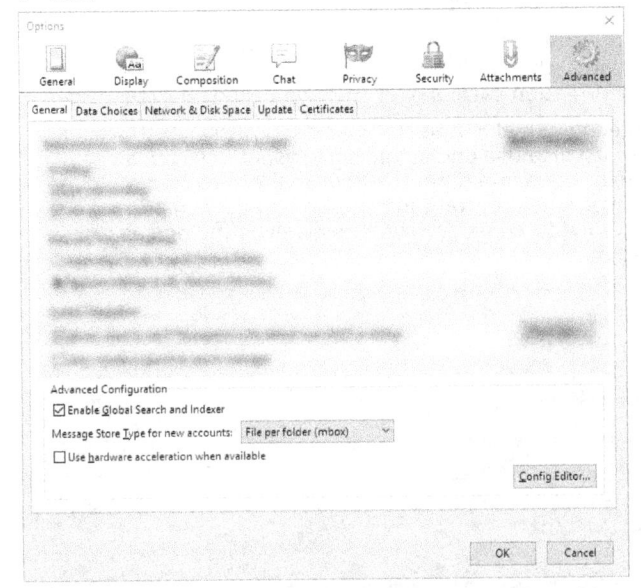

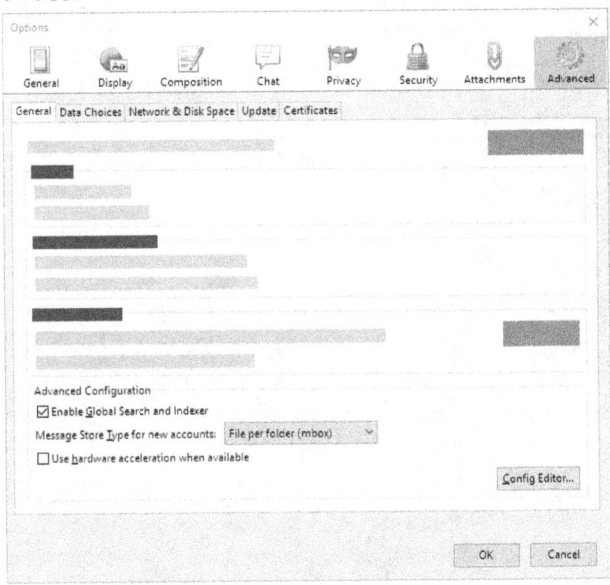

✔ **Yes:**

✔ **Yes:**

Images of software

Consider capturing a sequence of windows

If users need to perform an action that opens a new window, you may want to show the relationship between the original window and the new window. Or you may want to show how information carries over from one window to the other.

Instead of adding multiple individual images to your document (which would need a lot of space), you can also use one single screenshot in which all of these windows are visible at the same time. However, take care that the overlapping windows don't cover any important information. In particular, make sure that the title of each window remains visible, as well as the places where you've clicked or entered data.

Example:

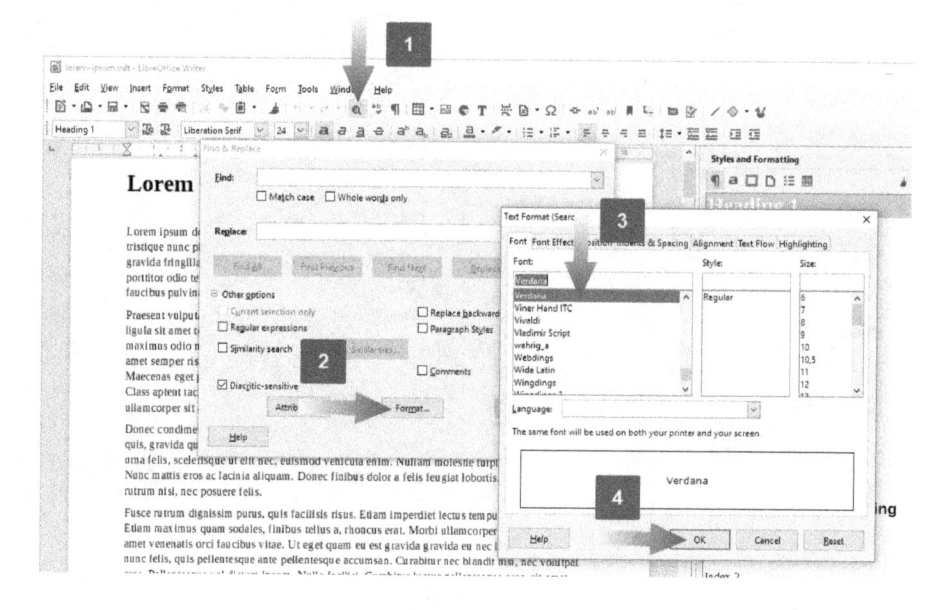

Related rules

When to include the mouse pointer? 183

Optimize each screenshot for its particular purpose 207

2.5.4 When to include the mouse pointer?

> In general, don't show the mouse pointer on your screenshots.
>
> More often than not, the position is just arbitrary and the mouse pointer is nothing but unnecessary clutter in the image.
>
> Only show the mouse pointer for indicating where an action needs to be performed.

When a mouse pointer in an image is helpful

Showing the mouse pointer can be an intuitive way to direct the users' eyes to the place where they need to act.

- When presenting conceptual or reference information, usually there's no need for pointing out some action. Here, it doesn't make sense to include the mouse pointer into the image. It would only add clutter but no value.
- When describing tasks, displaying the mouse pointer can be helpful to indicate the place where users need to act. However, in images that show the result of an action, the mouse pointer is not necessary.

Where to place the mouse pointer

Best, place the mouse pointer directly on top of the control that needs to be clicked. However, take care not to cover any symbols or text (or at least cover as little of them as possible so that they remain completely recognizable).

Often, the optimum position for the mouse pointer isn't right in the middle of the control but slightly shifted towards the bottom right:

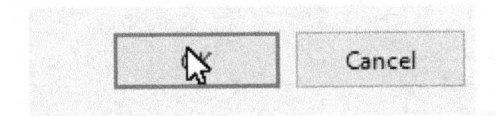

✖ **No:**

Images of software

✔ **Yes:**

[OK] [Cancel]

Tip:
If the mouse pointer turns into an I-beam pointer where users need to click, it's often better to place the mouse slightly below. The I-beam pointer may be hardly visible in the image.

Improving the visibility of the mouse pointer

Within screenshots of large windows, the standard mouse pointer can be quite small and hard to find.

Most screen capturing tools provide an option of capturing the image without the mouse pointer and can then add an enlarged or highlighted simulated mouse pointer on top of the image to attract more attention.

If you can, use this feature. It has several advantages:

- The simulated mouse pointer is more clearly visible.
- You can take as much time as you need to find the exact best position for the mouse pointer, and you can edit the position at any time in the future.
- You can later remove the mouse pointer from your image if you need to, for example, if you want to reuse the same image for another purpose.

Images of software

✘ No:

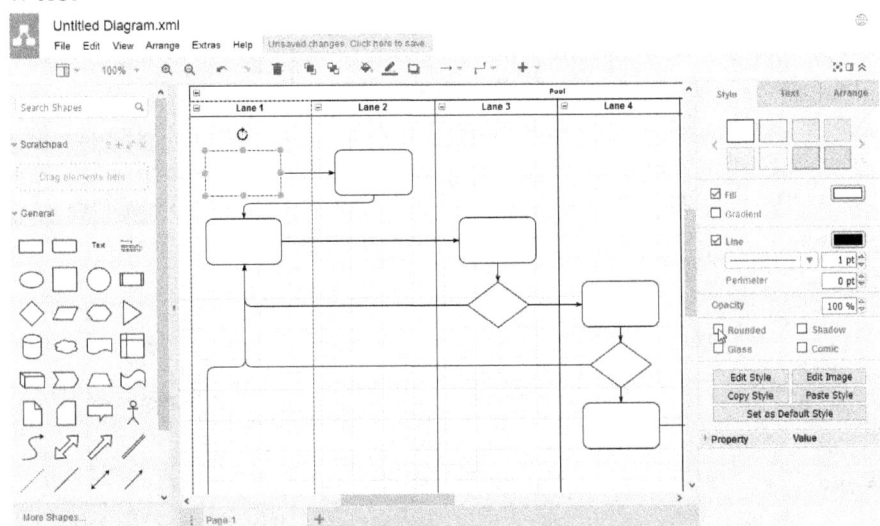

✔ Yes:

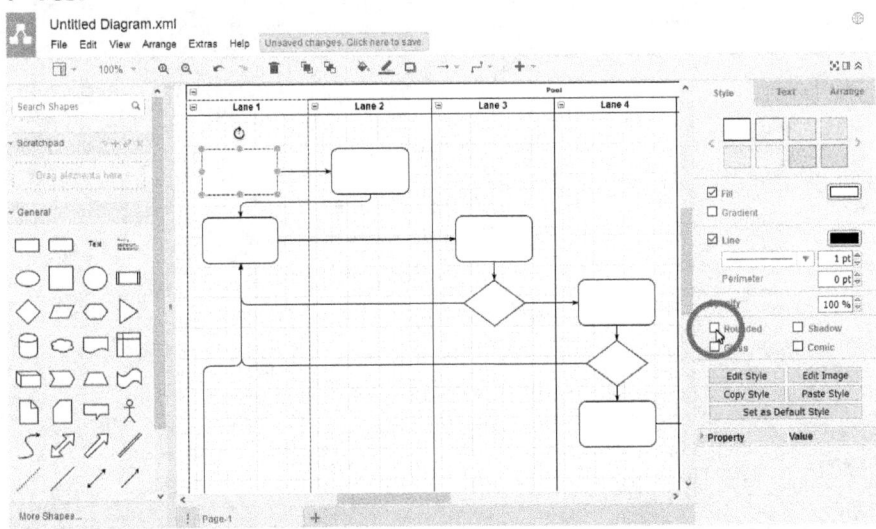

2.5.5 Don't maximize the windows to capture

When capturing an application or a window of an application, don't maximize the window to the full size of your display.

Rather, reduce the window to a size that:

- is as small as possible
- is large enough so that all elements that need to be visible are visible
- is large enough so that the window still looks natural with all elements in their typical positions

Example

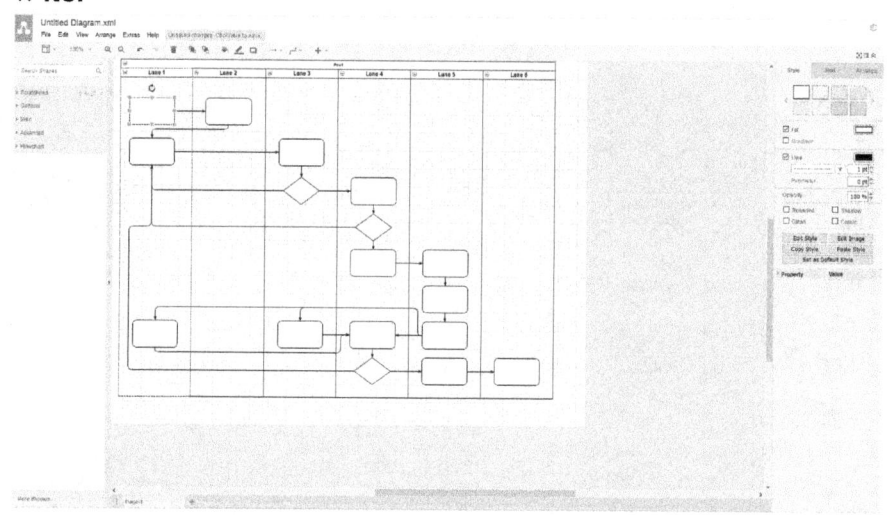

✖ **No:**

Images of software

✗ No:

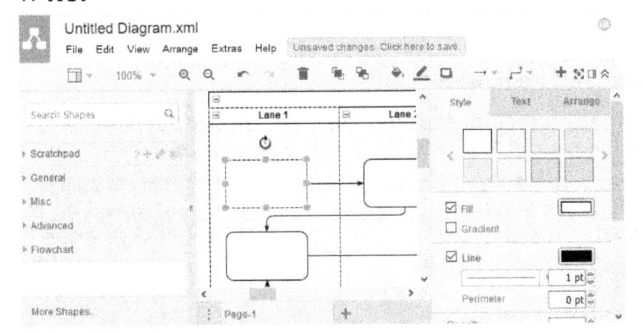

✓ Yes:

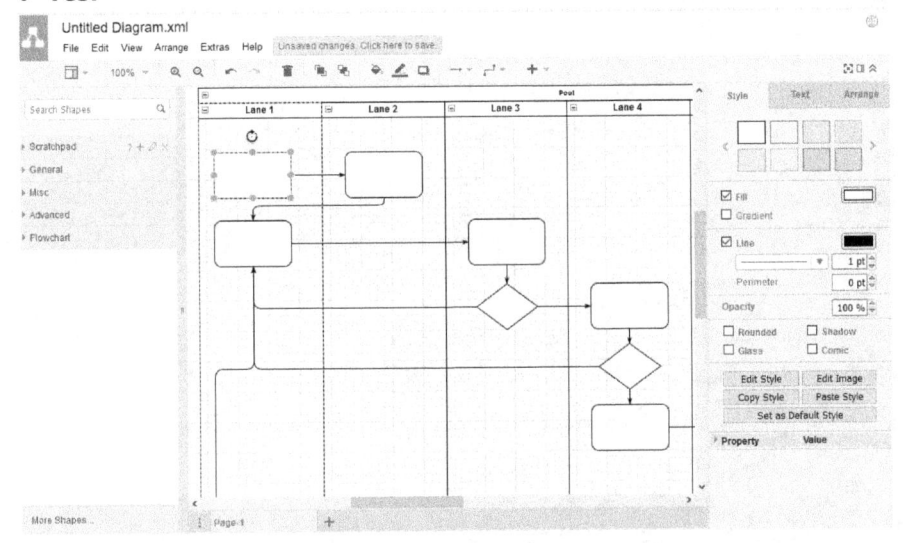

Why smaller images are often better

Smaller images have various advantages compared to those captured at maximum size:

- Small screenshots don't waste space on screen or paper.
- Small screenshots don't contain much superfluous information.
- In a small screenshot, the components within the window are closer together. The users' eyes don't need to move. It's possible to grasp the whole image at one glance.

Images of software

- Small screenshots don't need to be scaled down. So text within the image remains more legible.

What to do if a window isn't resizable

Some windows don't have any indicator that they can be resized but when you hover the mouse pointer over a window's borders, you actually can.

In case a window really has a fixed size, it mostly doesn't make sense to resize the window anyway because the window's content isn't responsive. However, if there's much space unused in the window, you can still tweak the image *after* shooting (see *Get rid of dead space* 189).

Related rules

Get rid of dead space 189

Avoid confusion with the real UI 193

Use standard settings 194

With web applications, don't show the browser 197

Show meaningful data 199

Hide private data 202

188

2.5.6   Get rid of dead space

> Sometimes, a software user interface contains large sections that either:
> - consist of nothing but white space
> - contain very uniform data, such as a list of names or addresses
>
> Showing a full screenshot of such an interface results in a low density of information. It essentially wastes space on paper or screen.
>
> Avoid having such dead space in your screenshots.

Consider resizing the window

Often, the best way to avoid having dead space within a screenshot is to resize the window that you're going to capture.

- Make the window small enough so that areas that don't need to be visible disappear.
- However, also make the window large enough to retain its typical visual structure. The window should still look natural.

Tip:
Some windows don't have any indicator that they are resizeable, but in fact they are.

Images of software

✘ No:

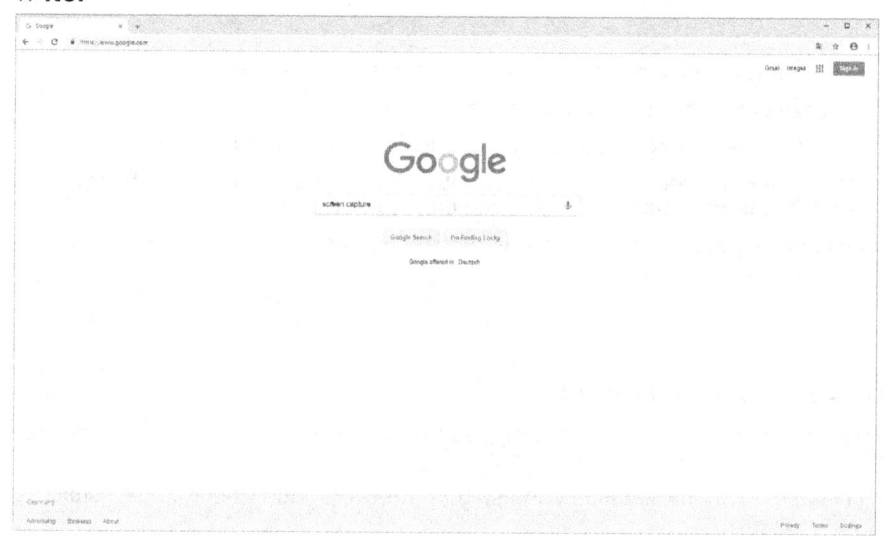

✔ Yes:

Consider cropping one or two sides of the image

If a window isn't resizeable, or if resizing doesn't remove the dead space, or if you've forgotten to resize the window, cropping the image at one or two sides can be an option.

- Only crop the image if this doesn't also remove any important elements that users should still be able to see.
- Make sure that it's visually obvious that the image has been cropped (see *What to show in a screenshot?* 177).

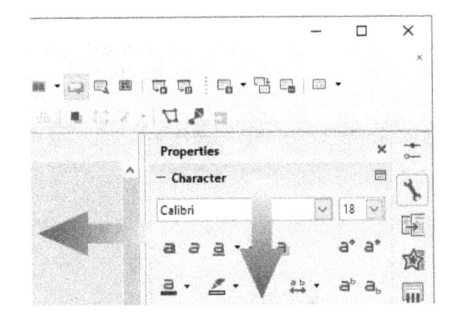

Consider "sliding together" the image

If the dead space is within the center area of your image, it's often possible to crop out the dead space by sliding together the upper and lower section or the left and right section of the image.

The result is a fake image that has less dead space than the original has. Users typically won't notice this manipulation. But even if they do, it's uncritical.

✘ No:

✔ Yes:

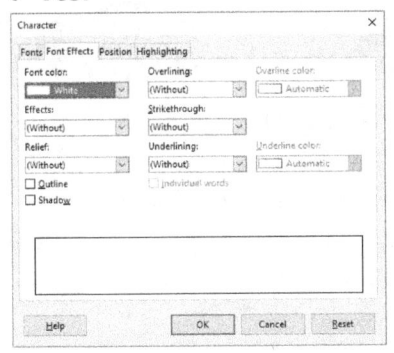

Images of software

If it's important that users *do* understand that the image has been cropped, instead of sliding both sections completely together, you can add some visual cue, such as two faded or torn edges.

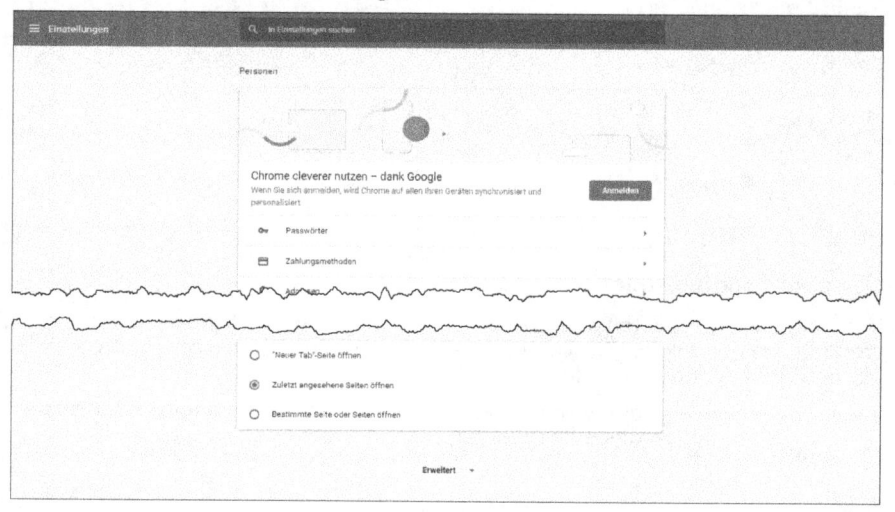

Related rules

Don't maximize the windows to capture 186
Avoid confusion with the real UI 193
Use standard settings 194
With web applications, don't show the browser 197
Show meaningful data 199
Hide private data 202

2.5.7 Avoid confusion with the real UI

When looking at screenshots in online help, users sometimes mistake the image for the application and end up clicking on the screenshot rather than on the actual program.

This isn't because users are stupid, but it happens when the screenshots look exactly like the software on their display.

To avoid this problem, make sure that your screenshots clearly differ from the software user interface in at least one characteristic.

Ways of making a screenshot look other than the software

You can:

- Make the image a bit smaller than 100% (but still fully legible).
- Use a clickable thumbnail of the image that opens the full-size image only in a lightbox.
- Crop the image (best add some indicator that the image has been cropped—see *What to show in a screenshot?* 177).
- Include some sample data.
- Add some highlighting on the image (see *Emphasize what's important* 77).
- Simplify parts of the image (SUI—see *Use a real screenshot or an illustration?* 174).

> **Related rules**
>
> *Don't maximize the windows to capture* 186
> *Get rid of dead space* 189
> *Use standard settings* 194
> *With web applications, don't show the browser* 197
> *Show meaningful data* 199
> *Hide private data* 202

2.5.8 Use standard settings

To build trust and to avoid confusion, what a screenshot shows shouldn't differ from what users see on their own screen.

Most operating systems let users adapt many settings, such as window colors and font sizes. Also, many applications let users choose from various skins, toolbars, etc. You can hardly match each setting, but you can maximize the likelihood that your screenshots match as many settings as possible.

Before taking your screenshots:

- Make sure that you're using the default settings of your operating system.
- For web applications also make sure that you're using the default settings of the browser.
- Make sure that you're using the default settings of the software that you're documenting.

Check the settings of your operating system

- Turn off any possible night mode.
- Don't use any custom scaling.
- Use the default color scheme.
- Don't use any custom fonts.
- Use standard font sizes.
- Remove any custom desktop background images.
- Clean up your desktop.
- Clean up your taskbar.

If the software to be documented is a web application, also check the settings of your browser

With many screenshots, it's best not to show the browser window at all but only what's inside (see *With web applications, don't show the browser* 197). However, some settings may even affect this area.

- Make sure that the zoom level is set to 100%.
- Use the browser's default theme.
- Use the browser's default settings for the address bar, for the toolbars, etc.
- Disable any browser add-ons / extensions / plugins that add visible elements to the toolbar or to other places.

Check the settings of your software

- Use the default installation and default configuration.
- Reset all settings back to their defaults.

Note:
Resetting all program settings to their defaults can sometimes be tricky. Many programs remember their settings even after uninstallation and reinstallation.

Mind the locale

Be aware of the fact that various standard UI elements and windows are actually supplied by the operating system and shown in its language even if it looks like they are part of an application.

If, for instance, you run some English software on German Windows, you might end up with something like this:

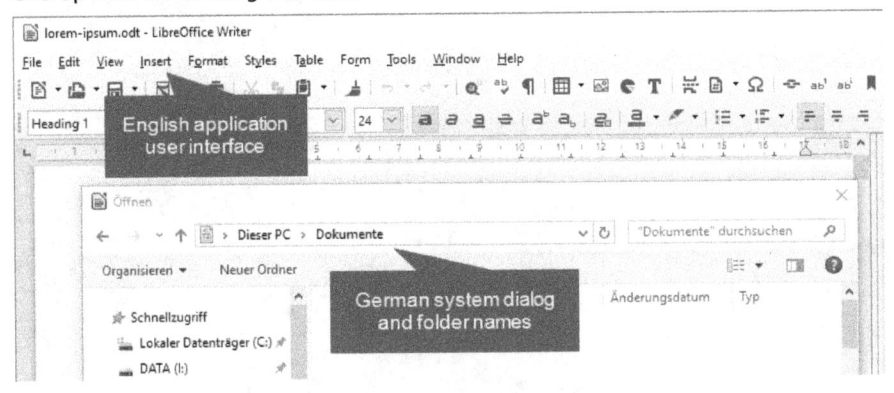

Therefore, on your operating system (and for web applications also in your browser as well) always use the same language and locale that you show in the screenshot.

> **Related rules**
>
> Don't maximize the windows to capture 186
> Get rid of dead space 189
> Avoid confusion with the real UI 193
> With web applications, don't show the browser 197

Images of software

Show meaningful data 199
Hide private data 202

2.5.9 With web applications, don't show the browser

If you take a screenshot of a web application, it's usually best to capture only the application itself.

Don't include the surrounding browser window with its frame, title bar, address bar, toolbar, and status bar. These elements aren't needed for orientation and aren't part of the application. So they don't add any value to the image either.

Example

✘ No:

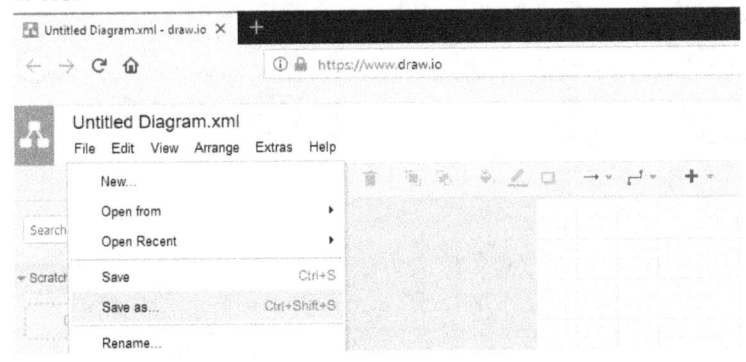

✔ Yes:

Images of software

Advantages of not showing the browser window

- With web applications, users use different browsers on different operating systems on different devices. Showing one specific system and browser may confuse users who are using other systems and browsers.
- As soon as there's a new version of the browser with a changed UI, if your screenshots show the browser, they will look outdated. So you may need to update your screenshots even if your web application hasn't changed.
- Showing the browser window needs more space without adding any usable information.
- Showing the browser window makes the image more crowded without adding any usable information.

> **Related rules**
>
> *Don't maximize the windows to capture* 186
> *Get rid of dead space* 189
> *Avoid confusion with the real UI* 193
> *Use standard settings* 194
> *Show meaningful data* 199
> *Hide private data* 202

2.5.10 Show meaningful data

> Don't underestimate the importance of the data visible within your screenshots.
>
> Many users use this data for orientation and as examples.
>
> If these examples don't work, or if these examples aren't realistic and practical, this can seriously affect the credibility of your documentation.

Make it realistic

Use data that's as plausible as possible. This helps users to mentally transfer what they see to their own tasks.

Finding realistic applications and examples can be one of the most challenging tasks when documenting a product. If you're not a subject matter expert yourself, don't hesitate to ask one for assistance. Depending on the application that you're documenting, collecting reasonable data and getting it into the application can be quite time-consuming and may require early planning.

In case you're sharing your installation and data with other persons, make sure that these persons don't accidentally delete your data or add any data that you don't want to publish. In particular, tell everybody involved not to add any confidential data and no data that might not be politically correct (see also *Hide private data* 202).

Keep it simple

The best examples are those that are simple enough so that users can easily identify how things are organized and how things work together.

Often, it's enough to have only some simple prototypical data rather than a wealth of complex data.

Don't fill in data where you don't need to

Don't feel obliged to enter sample data into every screenshot and into every field. Only enter sample data if this adds any value, which is typically the case if:

- the data is an essential part of what the image illustrates (example: showing demo names in a window that has been designed for selecting a person)
- the data provides inspiration on how to use the software
- the data shows the syntax required for entering information

Images of software

- the data shows the correct result of a described action

In all other cases, *not* showing sample data where it *isn't* needed has several advantages:

- You save time taking the screenshot.
- You save time updating the screenshot in the future.
- You save time translating the screenshot.
- The image remains clear and uncluttered.

Use a consistent demo project if you can

If possible, base all screenshots on a common demo project.

This makes your documentation very consistent and helps users in understanding how things work together.

Match the images with the text

If the text of your documentation mentions any particular examples or data, make sure that the screenshots also show exactly this data.

Anything else would be most confusing.

Save your sample data for future updates

Depending on the product to be documented, entering meaningful sample data can be a lot of work. If you can recover this data when needed, this is a major time saver.

In particular, think ahead about future updates of the documentation. Many test and demo installations (plus its data) are deleted after some time by people who aren't aware that the data may still be needed for documentation.

For this reason, always make a backup copy of your sample project and store this backup copy along with your documentation source files.

Mind the gender and ethnic groups

Depending on the markets to which you sell your products:

- If your data contains names, use both male and female names, as well as names from various countries and ethnic groups.
- If your data contains any geographical address data, such as countries, cities, etc., vary them appropriately.

Consider using English sample data

Depending on the amount and complexity of data, it can take a lot of time to translate it into all languages of your documentation. Sometimes, the advantages don't outweigh these costs.

In this case, you should consider showing sample data only in English.

Related rules

Don't maximize the windows to capture 186
Get rid of dead space 189
▸ *Avoid confusion with the real UI* 193
Use standard settings 194
With web applications, don't show the browser 197
Hide private data 202

2.5.11 Hide private data

With each screenshot, be careful that it doesn't disclose any data that mustn't be published:

- confidential data, such as URLs of demo servers, license keys, passwords, names of clients, etc.
- personal data of people, such as phone numbers and email addresses
- your own personal things, such as the pictures of your children on your desktop, the popping up message, or the forgotten browser tab with the dating site that you've just visited

Caution: Don't underestimate the importance of this issue. Failures in protecting privacy can become very expensive. In a worst-case scenario, you would need to stop shipping your product and destroy already printed manuals.

Other data that should not be visible on your screenshots are dates and version numbers. So your images won't look outdated in the near future.

Frequent pitfalls

In particular, check your images for the following:

- **On the desktop:** images, icons, files, folders, the date, installed and running applications
- **In Explorer windows, file open dialogs, and file save dialogs:** files, folders, most recently used files, dates
- **On the taskbar:** running applications, date
- **In browser windows:** history, bookmarks (favorites), URLs, web sites open on other tabs in the background
- **In the application:** passwords, names, email addresses, physical addresses, phone numbers, dates, version numbers, debugging information, client data
- **Everywhere:** things shining through transparent windows, notification popups (for example, from messengers), copyrighted images

Images of software

✘ No:

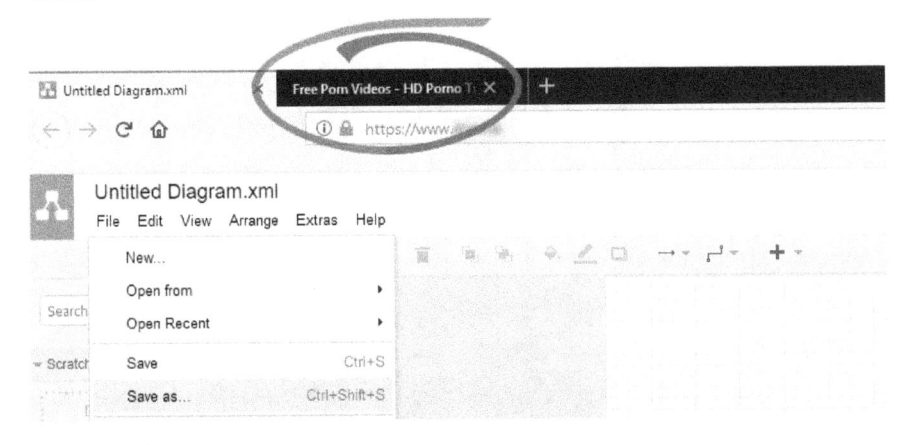

[screenshot showing a browser with a suspicious tab titled "Free Porn Videos - HD Porno T..." circled, alongside an Untitled Diagram.xml tab with File menu open showing New, Open from, Open Recent, Save (Ctrl+S), Save as... (Ctrl+Shift+S)]

How to handle sample data

Don't reveal any real client data in your screenshots. Instead, use fictitious sample data (see also *Show meaningful data* 199).

Tip:
Be careful with using existing test data. Some test engineers have a warped sense of humor and use names that shouldn't appear in a manual.

Sometimes, you can't remove confidential data for whatever reason, so you must make it unreadable.

> ⚠ **Caution:** Blurring is NOT a safe method of making confidential data unreadable! Unlike with arbitrary image content, reconstructing blurred text and numbers can be quite simple due to the letters' and numbers' repetitive structures. Better, black-out the text by putting some solid color on top of it, such as a filled rectangle.

Related rules

Don't maximize the windows to capture 186
Get rid of dead space 189
Avoid confusion with the real UI 193
Use standard settings 194
With web applications, don't show the browser 197

203

Images of software

- *Show meaningful data* [199]

2.5.12 If it saves you time, fake a screenshot

Sometimes, it can be difficult to take or retake a particular screenshot. For example, you might need to set up a special user account or to create a complex project just to be able to open a certain window.

If it saves you time or trouble, get creative and feel free to fake the image.

When faking makes sense

Typically, it doesn't make sense to fake an entire screenshot. But there are many reasons why you might like to fake some details of an existing one. For example:

- There's a planned change in the software user interface but it has not yet been implemented.
- There's a bug that makes it impossible to produce certain data or a particular state of the software.
- There's a bug visible in the user interface of the software.
- Sample data isn't available, and entering sample data would require lots of time.
- There's some confidential or personal data visible on an existing screenshot. However, taking a new screenshot isn't possible.
- There's a mistake on an existing screenshot. However, taking a new screenshot would take longer than editing the existing one.

Faking in case of software updates

A situation where faking screenshots can save particularly much time is when there have been only minor changes in the software.

Instead of reentering all sample data into the new version of the software, you may proceed as follows:

1. Take a screenshot of the new software version and open this screenshot in a pixel-based image editor.
2. Open the old, outdated screenshot in another window of the same image editor.
3. Either copy and paste the sample data from the old screenshot into the new one, or copy the new elements from the new screenshot into the old one—whichever is easier.

Images of software

> **Related rules**
> *Tips for taking screenshots* 209

2.5.13 Optimize each screenshot for its particular purpose

> In technical documentation, screenshots serve very different purposes. Don't design all screenshots alike but optimize each one for its particular use.

Screenshots that help to develop a mental model

These screenshots need to show many things at the same time in a very clear and simple way.

- Capture an area that's large enough so that it becomes clear how things work together. Often, you need to show a full window or even the complete application.
- Add some highlighting and comments on top of the image to draw attention to those parts of the user interface that are important for understanding.
- Consider using a series of screenshots instead of just one, which can better catch the logical flow or progression of windows. You may also be able to show several overlapping windows at the same time in one overall picture.

Screenshots that help to identify the controls mentioned in the text

These screenshots need to provide good orientation.

- Make sure that the selected area shows enough context (that is enough characteristic surrounding elements) so that it's clear what exactly can be seen in the image and where this is located within the user interface.
- Add some highlighting to guide the users' eyes to the relevant controls, such as a zoom box, a magnifying glass, a colored box, or an arrow. In procedures, focus on the action. Here, the most intuitive visual guidance is the mouse pointer (see *When to include the mouse pointer?* 183).

Screenshots that help to determine the states of the system and to verify results

These screenshots need to be uncluttered and highly focused.

- Show only the relevant area or even just a single control.
- Make sure that the image is large enough so that texts are easily readable and all other elements clearly visible.

207

Images of software

- As an option, consider putting some highlighting on the result or pointing to the result with an arrow.

> **Related rules**
> *What to show in a screenshot?* 177

2.5.14 Tips for taking screenshots

> Many online help systems and manuals contain a large number of screenshots. Even if you can save only a few minutes with taking, editing, and embedding each image, this can add up to quite an amount of time.
> - Use a professional screen capture tool.
> - Standardize your workflow.

Advantages of using a professional screen capture tool

In principle, you don't need a special tool for taking screen captures. You can:

1. Use the standard key combination of your operating system to copy the content of your screen into the clipboard.
2. Open any pixel-based image editor.
3. Paste the image from the clipboard into this editor.
4. Edit and save the image.

However, in many cases, this is not very efficient. Better, consider using a more professional tool that ...:

- ... can easily capture not only the whole screen or window with a single click but also individual panels and controls
- ... can also capture open menus, popup windows, and other elements that disappear when you press a button or click with the mouse
- ... other than general-purpose image editors provides automated functions for exactly those things that you typically need to perform with screenshots, such as adding torn or fading edges, blurring areas, adding a magnifier to a specific area, adding text callouts, adding an enlarged or highlighted mouse pointer, etc.
- ... can save the images in a fully editable format so that you can later change any callout texts, arrows, etc that you've added (some programs even let you undo cropping, which can be highly useful)
- ... can export callout texts into an XML file for translation
- ... lets you update an image just by exchanging the background image (the real screenshot), preserving all elements that you've added on top (such as arrows) and all other edits; this feature is also extremely helpful for translations

Images of software

Standardize file names

Set up and adhere to a consistent naming convention for the file names of your screenshots. This will later help you find a particular file or group of files when you need to:

- edit images
- delete images
- reuse images in another topic or in another document

A good naming convention indicates:

- the type of content
- the content itself

Example naming convention:

Prefix	Used for images of ...	Example
WIN-	Windows (both application windows and dialogs). Note: The window's title bar needs to be visible. Else, use a more specific prefix.	*WIN-program-options.png*
TAB-	Tabs or significant parts of tabs. Note: The title of the tab needs to be visible. Else, use a more specific prefix.	*TAB-address.png*
PAN-	Panels.	*PAN-project-explorer.png*
AREA-	Manually selected image areas.	*AREA-reports-list.png*
MENU-	Menus (main menus, submenus, context menus).	*MENU-file.png*
ICO-	Icons.	*ICO-open.png*
ILLU-	Illustrations (which means anything that's *not* a screenshot).	*ILLU-setup-process.svg*

If you save a backup copy of an original screenshot before editing the file, you can, for example, append a "0".

Example: *ICO-open-0.png*

Standardize storing locations

Similar to standardizing the file names, you should also use a convention as to where you save the image files.

Depending on the size and on the characteristics of your project, as well as on your authoring tools, you can follow one of the following approaches:

- You can save all screenshots to one common screenshot folder.
  This is a good approach if you don't have many screenshots in your project.

  Your folder structure can then look like this:

  |texts
  |images|screenshots
  |images|illustrations

- You can save the screenshots to the same folders where the texts of your documentation are stored. In addition, you can set up an extra folder for images that are used in multiple places of the documentation.
  This typically is a good approach for medium-sized projects.

  Your folder structure can then look like this:

  |setup
  |configuration
  |reporting
  ...
  |shared-images

- You can save the screenshots to a dedicated screenshot folder and add the same structure of subfolders here that you're also using for the text files. With this scenario, you can also set up extra folders for images that are used in multiple places of the documentation.
  This typically is a good approach for very large projects.

  Your folder structure can then look like this:

  texts|setup
  texts|configuration
  texts|reporting
  ...
  |images|setup|screenshots
  |images|setup|illustrations
  |images|configuration|screenshots
  |images|configuration|illustrations
  |images|reporting|screenshots
  |images|reporting|illustrations
  ...
  |images|shared|screenshots
  |images|shared|illustrations

Create a library of reusable objects

When editing your screenshots, you'll frequently need the same few basic elements and effects repeatedly. To standardize your images and to save time, it's a good practice to organize these elements into a small library.

Some screen capture tools already come with a special feature for setting up such a library. If your tool doesn't have this feature, you can create a master image to copy and paste from.

> **Important:** In either case, save and archive your library along with your project. So you or somebody else can access and reuse the library again in case the documentation and its screenshots need to be updated in the future.

Your library can include, for example:

- a highlighted mouse pointer
- a highlighting box
- an arrow for pointing at particular elements
- an arrow for indicating a movement
- a zooming effect
- a line
- a speech bubble or other element for callout text
- numbers for indicating steps or a sequence

Design your elements so that they use consistent colors, fonts, line widths, etc.

Make sure that all elements work as well when printed on a black-and-white printer.

Don't get pedantic with updates

Many software user interfaces change quite often. This is especially the case with early versions of a new software product. Some changes can affect a large number of images (example: a menu or the color scheme have changed). Constantly updating all screenshots, maybe even in multiple languages, can mean a lot of work.

If only the design changes to some extent, but not the functions, ask yourself whether it's actually necessary to update the screenshots or whether maybe you would be better off spending the time on some other enhancements of the documentation. Find the right balance between the two extremes:

- **Extreme 1:** You can update each screenshot so that it exactly matches the current user interface.

Images of software

This is the more user-friendly approach. It leaves users in no doubt that they are in the right place. Also, you don't run any risk of giving the impression that your documentation isn't up to date.

However, this approach can be expensive.

- **Extreme 2: You can update screenshots only if there has been a major functional change.**

    This saves you money. The savings multiply with the number of languages of your documentation. It also saves you time so that you can release the new version of your software and its updated documentation more quickly.

    However, some users may find the discrepancies between the software and the images annoying.

The optimum is somewhere in between. In the end, it depends on your audience how pedantic you *need to be*, and it depends on your budget how pedantic you *can be*. In general, users with a technical background and higher education are the most tolerant ones.

Related rules

What to show in a screenshot? 177

When to include the mouse pointer? 183

Don't maximize the windows to capture 186

▸ *Use standard settings* 194

With web applications, don't show the browser 197

Show meaningful data 199

Hide private data 202

If it saves you time, fake a screenshot 205

213

2.5.15 Tips for translating screenshots

Translating a large number of screenshots can be a lot of work. It multiplies with the number of languages that you need to provide.

- This can result in high costs.
- This can result in delayed time to market.

There are a number of ways of how you can optimize the efficiency of the process and find a compromise between cost and quality. But no matter how you proceed: Plan the translation process early so that you don't end up with unexpected delays that defer the shipment of your product.

Texts that may need translation

Often, the texts of technical documentation get translated by external translation agencies. Most authoring systems used for software documentation can export XML-based translation packages that the translators can then edit within their own translation systems. Finally, the translators send back the translation package, and it's re-imported into the authoring system.

For screenshots, translation is more difficult and involves several tiers. There can be three different kinds of text in a screenshot that may all need to be translated:

- the texts of the software user interface shown in the screenshots, such as the names of menus and buttons, labels, etc.
- any possible sample data visible in the screenshots, such as the names of objects that were added by users
- any texts that were added on top of the screenshots after they had been taken, such as callout texts

Translating the texts of the software user interface

These texts have already been translated when you take the screenshots. This translation is typically part of the software development process.

However, you need to define who is responsible for taking the screenshots:

- External translators mostly aren't qualified to use the documented software well enough.
- The documentation department typically has a time bottleneck when translations into multiple languages need to be handled at the same time.

    But there can also be a more severe problem: With unknown languages, operating the software and entering data may be almost impossible—especially when a language uses an unfamiliar alphabet. Also, you should set your operating system to the corresponding locale while taking the screen captures, which is another challenge.

- To resolve this problem, screenshots of translated software are often made in a national office of the manufacturer or by an employee who speaks the corresponding language.

Translating the sample data visible in the screenshots

Unlike the texts of the user interface, these texts have not yet been translated as part of the software development process.

To do so, you have the following options, which all have their pros and cons:

- If the screenshots in a particular language are going to be made by a national office of your company or by a colleague who speaks the language, you can ask these persons to translate the sample data as well.

  Usually, this is the easiest and fastest way.

- You can make a list of the texts and send them to a translator.

  This will typically yield good results, but it can take a lot of time.

- You can use English sample data.

  This clearly isn't the optimum solution for the users, but it may save you a lot of time and thus money. In particular, this can be a good approach for those languages:

  - that are spoken only by a small proportion of users
  - that are spoken in countries where many people have a good command of English

Translating the texts added on top of the screenshots

If you're using one of the more advanced tools for taking and editing your screenshots, you can export these texts to an XML file and then have them translated in a workflow similar to the translation of regular text. Some tools also use XML-based file formats even for storing the screen captures, which makes the export procedure unnecessary.

If your tool don't support this, you can write a list of the texts manually, send this list to a translator, and then copy and paste the translated texts back into the images.

Option: Reproducing the user interface as a vector image

It may sound weird, but some companies actually don't use any "real" screenshots at all. Rather, they use mock-ups made with some vector-based graphics software. This doesn't mean that everything has been drawn manually. Some

215

image editors can use a real screenshot as a starting point and then convert this screenshot with the help of a tracing function into editable vector objects. Also, many objects are standardized in a user interface and thus reusable.

Instead of reproducing the whole image as a vector image, another variant is to use an English screenshot (or whatever your primary language is) as a background and then to overlay this in an image editor with text fields that have a solid background color and thus cover the original texts.

The advantages of these approaches are:

- The images can be made, in every language, even before the software is final and has been translated.
- All texts are editable and thus translatable.
- The texts can easily be updated at any time.
- If you use SVG as a file format, in online documentation you can have the search function index the texts. So if users search for any term that is used in the user interface, they can find the corresponding image in the documentation.

The disadvantages of these approaches are:

- It takes much longer to create the image for the base language than it would take to capture a plain screenshot.
- The images may slightly differ from the actual user interface of the software.

As a conclusion, these approaches can be especially interesting if:

- you need to translate your images into many languages
- translated versions of the user interface are not yet available when you need to add the images to your documentation

Option: Adding editable text overlays in your authoring tool

Another option of "faking" a translated user interface works as follows:

1. In your authoring tool, instead of a screenshot add a table. Use a table that has multiple rows and multiple columns that all have a fixed width and height, for example, 5 mm.
2. To this table, add the screenshot as a background image.
3. Where you need to translate any text, merge the table cells that are right on top of this text. (If your target language runs wider than the original language, add a few extra table cells.)
4. To the merged table cell, as a background color assign the background color of the text behind the cell.
   The original text is now covered and thus invisible.

Images of software

5. To the merged table cell, add the translated text. Apply the same font, font style, and size that the original text has (create a special character style in your style sheet for this purpose).
   The translated text is now visible on your image at the same place where the original text was.

What can make this method interesting, is:

- The texts are part of the main content and can thus be translated along with the main content by a translation agency. It doesn't require any additional translation workflow.
- The texts are part of the main content and are thus automatically indexed by search—both in the authoring tool (useful to the author) and in the online help systems that the authoring tool produces (useful to the users).
- If your authoring tool supports variables, you can use them for both the labels visible in the images and also in the body text of the documentation. So these texts only need to be translated once and are automatically consistent. Of course, it means quite some work to add them, but the more languages you have the greater the benefit is.

Option: Not translating screenshots

Some companies don't translate their screenshots even though they *do* translate the documentation.

Case 1: If the documented software has *not* been translated into a particular language but the documentation has, the case is simple: Just leave the screenshots in their original language—typically English.

Example: A Program has an English user interface. There's no way of switching the UI language to Dutch, but anyway you provide a Dutch user manual. As the Dutch user manual describes the English software, all screenshots are also in English. Sample data visible in the screenshots are also not translated.

Case 2: If the documented software *has* been translated into the documentation's language, typically the screenshots are also translated. However, you may optionally also consider describing and showing the English user interface rather than the locale one. From a user's perspective, this is not the best solution. However, it takes less time to create and it costs less money.

In practice, there can be yet another reason for this approach: Often, there are plans of translating the user interface, but it lags back in time. So there's actually no other way than either documenting the English user interface or waiting for the translated one.

How well this actually works, largely depends on your audience and on their ability and willingness to understand the English user interface.

Sometimes, the best solution is a combined approach: Translate all images into those languages where you have many users and key clients. Use untranslated English images in the documentation in those languages where you only have a small number of users. Often, these are smaller countries with rare languages where, luckily, for exactly this reason users are used to the fact that many documents aren't translated into their language but are only available in English.

> **Important:** No matter how exactly you proceed: When not translating your images, make sure that you don't have any legal or contractual obligations to supply fully translated documentation.

2.6 Video design

Good videos inspire trust and communicate their message very clearly.

Avoid all unnecessary things. Use video to inform, not to impress. The following tips will help you to do so:

Which type of video? 220
Keep videos short 224
Provide navigation and orientation 226
Where to place the videos? 230
Use meaningful poster images and titles 233
Create a storyboard 235
Add text, closed captions, or voice-over? 240
How to handle warnings? 243
Show the presenter? 245
One or two presenters? 248
Male or female presenters? 250
Professional voice, own voice, or synthetic voice? 251
Use music? 253
Include ambient noise? 255
Keep the video simple 257
Use effects wisely 262
Consider embedding also still images 267
Consider linking to further information 268
Consider adding a call to action 270
Consider creating a brand 272
Create small video modules 274
Standardize your video modules 276

All general principles for visualization apply as well. Thus, also see:
Common basics of visualization 19

219

2.6.1 Which type of video?

> Before you design and create a video, clearly define your goals.
> Only the right type of video can convey your message optimally.

Basic types of videos

Video type	Characteristics
Teaser video	Briefly introduces a product or a function without going into the details of the operation.
	The main purpose can either be to sell the product or to encourage users to explore and use a particular feature ("sell" the feature).
	Thus, the focus of a teaser video is clearly on the benefits of using the product or function.
Product presentation video	Introduces a product in more detail than a teaser video does:
	• What can the product do?
	• How does it work?
	• How is it used?
	• What does it look like?
	• What's the user experience like?
	Even though these videos are typically used for marketing, they can also take a dual role and be used in user assistance as well.
	Therefore, a product presentation video should be designed to both entice and educate at the same time.
Instructional video (short how-to-video)	Shows how to accomplish a single, specific task.
	Clearly educational.
Utility video	Special form of instructional video.
	Each video is very short (just a few seconds) and demonstrates only one step of a procedure. The individual videos are linked to each other and

Video design

Video type	Characteristics
	only together add up to the complete instruction. Often, this type of video has no audio. The particular strength and use case of this type of video are long, complex instructions, such as assembly and maintenance instructions: - Users can remember and then perform each (short) step without needing to rewind. - Conditional branches are possible. - The videos can often be reused in various procedures and or for various product versions. (The probability that a short video can be used more than once is greater than with a long video.) - The videos are easy to update. (In case of a product change, only the short video that's affected needs to be updated but not one big video.)
Microvideo	Very short video of just a few seconds. Presents only one single process or idea. Often doesn't have any narration or text but only shows the action or a short animation.
Tutorial (long how-to-video)	Describes complex tasks that may even involve several devices or programs, and shows how these items work together. Typically works with an example, and provides the knowledge to apply the example to other applications.
Screencast	Generic term for all sorts of software videos, regardless of their particular type and purpose.
Explainer video	Typically uses simple animation (motion graphics) or stop motion video in combination with paper figures, plastic figures, or similar objects. Despite its name, this kind of video is most often used in marketing for teaser videos. But it can also be an entertaining and easy form to explain a concept.

Video design

Video type	Characteristics
	However, because of their high degree of abstraction, explainer videos are not good for showing actions.
Conversational video	Uses a conversational tone to project authenticity and transparency. The presenter typically is a subject matter expert—not a professional actor or voice over artist.
	Often, the video shows the presenter at least briefly to build some form of personal relationship.
	The content may be a product presentation or an instruction.
Recorded webinar	Can be used for longer tutorials or even for product presentations aimed at particularly interested users.
	Depending on the presenter, the webinar can convey a high degree of authenticity and competence with simple means.
	If the webinar is very long and has a clear structure, it may be split into several videos.
Other ...	Other types of product videos, which are normally not used in technical documentation, include: • image videos • interviews • unboxing videos

Real video or animation?

Like an image, a video doesn't necessarily need to show real, physical objects. Instead of showing a movie that was shot with a camera, you can also create movies that consist of animated scenes. Sometimes, a real movie is also merged with an animation that's projected over the images taken with a camera.

Advantages of real video
- Shows exactly what the product or action looks like. This minimizes the risk of misunderstandings.
- Shows parts in their exact positions and perspective.

- Users don't need to perform any mental transfer, which makes viewing the video and imitating actions easy.
- Because it's more realistic than an animation, it's more memorable.

Advantages of animation
- Can simplify something complex by reducing it to its key elements.
- Can show even things that aren't visible in reality (example: the inside of the cylinder of a running combustion engine).
- Can show even very large objects that cannot easily be captured with a camera.
- Can already be created while the product is still under development or if the product is not available for shooting video.
- Depending on how an animation was created, it can be updated to some extent when the product changes.
- Text visible on objects can be translated.

Effort and cost

Animation doesn't necessarily need to be more expensive than real video.

Sometimes, existing CAD data can be used to render an animation with very little extra effort.

Also, mostly an animation doesn't need to be sophisticated to be effective. Unless the style of your video is very formal, you can, for example, create a simple stop-motion-video with the help of some cardboard or paper objects, plastic figures or whatever can serve as a good symbol for what you need to demonstrate.

Related rules

Interactive video (hypervideo) 316

2.6.2 Keep videos short

There's no standard length or minimum or maximum length of a video. What's the best length always depends on the subject and on the attention span of your audience.

However, given some particular information that a video is to convey, the shorter the video the better. Users watch instructional videos for getting their jobs done—*not* for having a good time. A shorter video saves users time. Other than with printed information, skipping irrelevant content is very difficult in a video.

So get to the point quickly. Leave out all things that are irrelevant in the context of the video.

Advantages of short videos

Compared to longer videos, short videos:

- are more likely to be watched to the end
- are less costly to produce
- are less costly to translate
- are less likely to need an update in case the product changes
- are more likely to be able to be reusable in various places

With complex subjects, divide a video into several individual videos. So users can view individual content in a more targeted manner. Users typically have a very specific question and aren't willing to watch a long video that spans multiple topics.

Disadvantages of short videos

Given some particular information, the shorter a video the better. A short video is always preferable, as long as the presentation quality doesn't suffer and the video doesn't run too fast (see also *In your videos, take your time* 305).

As to the question of whether it's better to have multiple short videos rather than one long video, a downside of having short videos is that those (few) users who are willing to invest more time into learning, need to find and start multiple videos. Also, you need to handle a larger number of video files.

Video design

Recommended lengths

In *general*, users prefer videos of approximately the following maximum lengths:

- up to 3 minutes for promotional videos
- up to 5 minutes for informational videos (depending on the content also up to 15 minutes)
- up to 15 minutes for tutorials (provided that the tutorial gives its users an obvious benefit)

However, when embedded into technical documentation, videos should be much shorter. The shorter you can make it, the better.

> **Related rules**
> *In your videos, take your time* 305

225

2.6.3 Provide navigation and orientation

> Videos cannot be skimmed as quickly as texts and still images.
>
> If a user seeks particular information, it may take (waste) quite some time until the video gets there. This "sluggishness" is the most important disadvantage of video.
>
> While the video is playing, the users are kind of trapped in there. Unlike with text and images, they only see what's currently playing but not what was before and what will come after.
>
> To overcome this limitation as well as possible:
> - Create multiple entry points into the video.
> - Constantly inform users about their current position within the video.
> - Make skipping (as well as repeating) particular content easy.

Ways of providing navigation

The base for being able to watch content selectively is that the videos have cue points that subdivide the videos into distinct scenes. In a procedure, for example, each step can be such a scene.

Depending on the used authoring tools, video formats, and video player, you have various options:

- You can add a table of contents of all scenes. This lets users select any scene at any time.

    So users who need only some very particular piece of information can access this information directly without wasting time watching the whole video.

- You can add extra buttons that immediately jump to the next scene or repeat the current scene.

    So users can precisely navigate in small intervals. Irrelevant content can be skipped; difficult content can be watched again, beginning at just the right position.

Video design

Example:

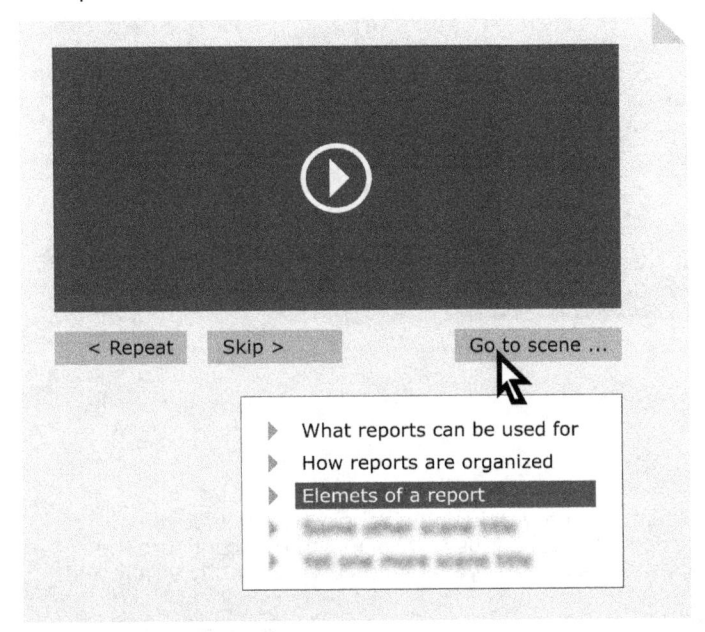

For procedures, a user-friendly way of embedding video can be to have it side by side with (or before) brief written instructions. In this case, each step is at least briefly mentioned in the text *and* shown in the video. This allows each user to use the medium that's best suited to their particular situation:

- video for initial information in the overall context
- text for selective information for more advanced users

At best, text and video are mutually linked:

- When a user clicks a step in the text, the video plays the scene that shows exactly this step. (Afterwards, the video automatically continues.)
- While the video is playing, the step that's currently being shown and explained is highlighted in the text.

With this approach, the description in the text works as a dynamic table of contents and progress indicator for the video. In contrast to a conventional procedure without video, the text can be very short here because the video provides the details if needed.

227

Video design

Example:

Changing the battery:

1. Remove back cover
2. Remove old battery
▶ 3. Clean connectors
4. Insert new battery
5. Reinsert back cover
6. Reset clock

Player controls

> ⓘ **Important:** Except for pure marketing videos, where in fact you *don't* want to make it particularly easy for users to skip content, configure the video player so that the available player controls are constantly visible and not hidden.

However, only enable those controls that are actually needed for your particular type of video:

Control	Recommendation
Play / Stop	Required. One toggling button is enough. An extra pause/resume button is typically not needed.
Progress bar with possibility to jump to any place within the video	Required except for very short videos (microvideo).
Display of time past Display of time remaining	Required except for very short videos (microvideo).
Navigation menu	Recommended for longer videos, especially if they show a complex procedure.
Go to next scene Repeat scene	Recommended for longer videos, especially if they show a complex procedure.

Video design

Control	Recommendation
Full screen	Recommended if the video's resolution is sufficiently high. Not recommend for low-resolution videos.
Show closed captions Hide closed captions	Recommended in case closed captions are available. (For details on when to use closed captions, see *Add text, closed captions, or voice-over?* 240.)
Change frame rate	Recommended only if bandwidth is an issue. Because your videos ought to be short, this should normally not be the case.
Change speed	Recommended only for longer videos and in case the video has no audio.
Mute / unmute audio Change volume	Not recommended. Users can use the corresponding controls of their hardware and operating system.

Don't start any video automatically

Always let the users decide whether they actually want to watch a video, and if so, when.

Do not run any video automatically. Imposing a video on someone is likely to cause annoyance.

When users need to start a video intentionally, they can take enough time to prepare both mentally and physically (take a deep breath, turn on the speakers, wait until the colleague across the desk has ended his or her telephone call, ...).

So always provide a button or symbol that users must deliberately click to start the video.

> **Related rules**
>
> *Use meaningful poster images and titles* 233

229

Video design

2.6.4 Where to place the videos?

Don't make a video something special. Place the video exactly where you would place a text or image with the same information if the video didn't exist.

In your documentation, don't set up a separate section with videos for their own sake. Nobody watches a video only because it's a video. In technical documentation, people primarily look for information—the medium (text, image, video) is secondary.

Best position for videos that explain a concept

In topics that explain a concept, include the video *after* you've first mentioned the concept in the text. Users can then decide whether they should invest the time into viewing the video or whether they can skip it.

If the text consists of multiple sections, put the video *between* an introduction and the explanation or example. This is particularly important if the explanation or example can only be understood with the help of the video. Thus, users first watch the (short) video, then read the elaborate explanation or example.

Best position for videos that show a procedure

If a video shows a complete procedure that consists of multiple steps, include the video *after you've introduced the procedure* and named any possible prerequisites.

If your text redundantly describes the procedure's steps as well, put this description below the video. Thus, users who prefer to read the text rather than to watch the video can skip the video but at least notice that it's there.)

If a video shows only one step of a procedure, include the video *after you've mentioned the step in the text*.

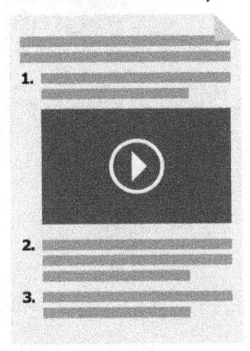

Preserve the visual structure of your topic

When embedding a video into a help topic, this doesn't mean that you need to embed it in its full size. Unless your video's dimensions are very small, this would disrupt the topic so that users would lose orientation.

You better can either:

- Embed the video into an expandable ("dropdown") section.

231

Video design

- Embed only a small thumbnail of the video's poster image and open the full video in a popup window or lightbox when a user clicks the thumbnail.

Using video on the welcome page

For audiences who are unlikely to read much text, best add a short video to the welcome page of your online help system. This can be a great way to pull users into the documentation.

This video should provide some brief onboarding information on the general operating principles of your product so that after watching the video, users can start exploring the product on their own. Also, this video can point out some particular help topics for more details.

2.6.5 Use meaningful poster images and titles

The number one reason why users abort watching instructional videos is not because the videos are too long, boring, or of poor quality.

Videos are mainly aborted because they don't cover what their users thought they would cover. This is the worst thing that can happen! Wasting time, severely frustrates users—no matter how good your videos are. Once one or two videos didn't deliver what was expected, users won't watch *any* of your other videos. You've then lost the game forever!

Help users find out what a video is about even before they click the start button.

Finding the right poster image

The poster image is the placeholder image that's displayed until the user starts a video (usually by clicking the poster image). By default, the poster image is the first frame in the video. However, many authoring tools also let you select another image from within the video or from somewhere else. Depending on how the video is embedded, the image may also be reduced to a thumbnail image.

A good poster image:

- is simple so that it can easily be recognized even if it has just the size of a thumbnail
- provides a meaningful preview of the content; for example, in a procedure, the poster image already indicates the first step of the action
- is superimposed by an icon that indicates that clicking the image starts a video

Finding the right title

Finding a good video title is very much like finding a good topic title. The title needs to communicate:

- The topic: What exactly is the video about?
- The information type: Does the video explain a concept, or does it provide instructions on how to perform an action?
- The level of complexity: Is the video a simple video for beginners, or is it a more demanding one for advanced users?
- The length of the video: What does it "cost" the user to watch the video in terms of time?

233

Video design

Don't try to advertize your videos to everybody. Users won't be amused if you steal their time by forcing a video on them that doesn't help them. However, *do* communicate the benefits of watching the video. This is an important and legitimate motivation for winning and maintaining the users' attention.

If your video explains a procedure, use a verb in the title. When users have a specific goal in mind, they will look for a verb in combination with a noun. The better you can anticipate these terms, the more often your video will be viewed. This is also important for search engine optimization if you publish your videos on the Internet.

Example titles

✘ **No:** *Principle*

✔ **Yes:** *Understanding the measurement principle (1:27 min)*

✘ **No:** *Templates*

✔ **Yes:** *Using templates to save time (3:45 min)*

✘ **No:** *Product Maintenance*

✔ **Yes:** *Performing weekly maintenance (7:45 min)*

Using video titles for linking

When linking to a video, best use the video title for the link anchor. If you've chosen a good video title, this is the text that best describes what the video will deliver.

In addition, include a video icon or the word "Video" into the link text. Else, users might expect to open a normal page when clicking the link.

✘ **No:** *Performing weekly maintenance (7:45 min)*

✔ **Yes:** *Video: Performing weekly maintenance (7:45 min)*

✔ **Top:** ▶ *Video: Performing weekly maintenance (7:45 min)*

> **Related rules**
>
> *Provide navigation and orientation* 226

234

2.6.6 Create a storyboard

> A **script** includes only the text that's going to be spoken in the video. A **storyboard** combines the script with additional information about what's to be shown visually, and which effects are to be used.
>
> Taking the time to write a storyboard for your video will ultimately *save* time overall.
>
> Your storyboard doesn't need to be fancy. The important thing is that you *do* have one at all.
>
> Having a storyboard doesn't mean that you must adhere to it strictly. Use it as a tool, but don't become its slave. Understand the storyboard as a part of an iterative development process.

Do you really need a script or storyboard?

Developing a storyboard is not a waste of time but streamlines your development process and increases the quality of your videos.

- If you have a storyboard, you can review and improve your content before spending lots of time creating the video. You can review the storyboard yourself, and you can have it approved by others before shooting the video (subject matter experts, the legal department, your boss, or your client in case you're a contractor). Editing a storyboard is always faster and cheaper than editing a video that has already been recorded.
- Based on the storyboard, you can prepare and rehearse all actions prior to the recording. This not only lets you perform the actions more smoothly. It can also identify possible problems, such as missing objects, data or files, insufficient user permissions, etc.

    Tip:
    Before starting to rehearse, ensure that you can reset everything to the initial state. When creating software videos, for example, back up all files.
- You can use the script text also for translation and for adding closed captions.

In addition, a storyboard also makes it possible to have someone else (such as an external service provider) create the video.

In case you don't have any time for preparing a storyboard, at least set up a rough outline of what you're planning to show, including the necessary processes and their steps.

If a video is going to demonstrate a task, the storyboard typically needs to be less precise and elaborate than with a video that describes a concept. This is because with tasks, the steps are already largely determined by the product. A video that explains an abstract concept needs a lot more creativity.

Storyboard design

There's no common standard for the design of a storyboard. Feel free to lay out your storyboard according to the specific requirements of your particular project and according to your personal preferences.

You can organize your storyboard either:

- as one long (multipage) table in which each row describes one scene
- in the form of a presentation where each slide describes one scene

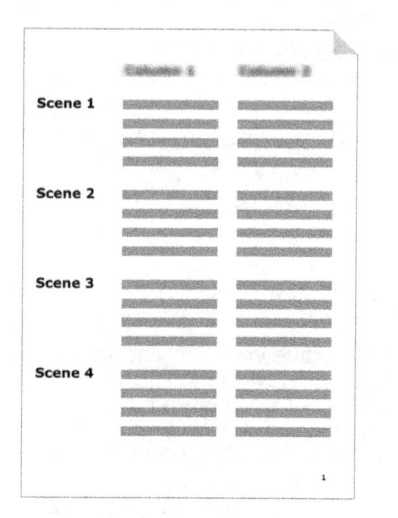

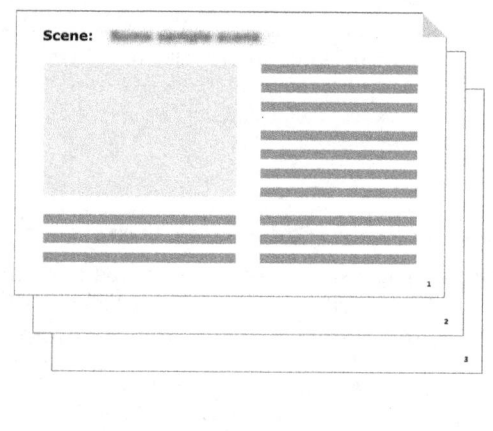

Typically:

- go with the tabular design for rather simple storyboards
- use a slide-based design if a storyboard needs to be more elaborate

Use telegram style for the descriptions in your storyboard.

However, in the script of the narration, specify exactly word for word what the presenter or narrator are going to say.

If your video includes any written texts, such as captions or annotations, include the exact full text as well.

Columns or sections to include into the storyboard

Keep your storyboard as simple as possible. Depending on the type of video that you're going to create, the storyboard may contain a selection of the following standard columns or sections:

- **Scene title:** Describes what's presented. If you're going to provide some scene navigation, the scene title can later also be used as the caption for the links or menu to access the scene.
- **Action:** Describes what's shown and done. Photos or screenshots can be helpful here but are optional. Text plus rough, hand-drawn sketches are perfectly OK.
- **Visuals / Effects:** Describes all types of modifications and elements that aren't part of the core video but that are typically added in a post-production process. For example, overlays, zooming and panning, all sorts of highlighting effects, texts added, etc.
- **Actors' voice:** Text that the person or persons are going to say who perform the actions in the video.
- **Narrator's voice:** Text that an extra narrator is going to say who isn't visible in the video.
- **Music:** Music played in the background.
- **CC:** Text of closed captions if your video is going to use any.

Additional columns can be, for example:

- **Location** (for hardware videos) or **Window** (for software videos)
- **Preparation** (list of things that need to be done so that the scenes can be recorded)
- **Timing** (start / stop / duration)
- **File name** of the video file to be created
- **Links** to other videos or to external URLs or files
- **Footage and files** to include into the video, such as existing images, logos, audio files
- **Interactive elements** (in case you're producing hypervideo)
- **Persons responsible** for particular tasks
- **Internal references**, such as author, date, status

> ℹ **Important:** Only use those columns that you really need. Sometimes, it can also make sense to combine two columns into one.

Video design

Feel free to start simply

When developing a storyboard, it can be helpful to start very informally. Just take your favorite word processor (or a piece of paper), and put down what you'd like to show and say.

Imagine you were explaining the process to someone from your audience. Don't care about spelling, grammar, word choice, making perfect sentences, and the beauty of your sketches. Focus on the content.

Later, you can refine this very first draft, but even then don't forget that the storyboard is just a temporary tool, not the final work.

Use spoken language

When scripting the narration of your video, best write it down while speaking it yourself. After all, it's going to be *spoken* text, not written text!

Tip:
Perform a so-called *table read*: Read your text aloud (if several presenters are to be involved have each person read their role). Listen and ask yourself: Is this actually the way you would speak to a normal person of your audience? Is this actually the tone of the company that you're working for? Is the company tone actually that formal? If not, edit the script. Also, this is a great chance for everybody involved to give the script some personality and thus ultimately to make the video more authentic (see also *Rethink you idea of "good"* [280]).

Optimize the script for easy reading

If your storyboard contains any text to be spoken by a presenter or narrator, format this text in a way that makes reading and emphasizing it easy:

- Use a short line length.
- Break the lines where there are to be little pauses of speech.
- Print words that are to be emphasized in bold or in color, or underline them.
- Add hints for the pronunciation of difficult names and words. Bear in mind that a professional voice-over artist may never have heard particular technical terms before. Mispronouncing them would ruin the credibility of the entire video.

✘ **No:** According to Professor Hu Jintao of Qinghua University, this is a demo of a poorly formatted script. As you can see, it's quite difficult to both read and to pronounce each sentence correctly.

✔ **Yes:** According to Professor [hoo jeen tow] (Hu Jintao)

of [ching hua] (Qinghua) University,

this is a demo

of a well-formatted script.
As you can see,
it's quite easy
to both read
and to pronounce
each sentence correctly.

Have your script or storyboard reviewed before shooting

Have a subject matter expert or a second person review your script before shooting the video.

In addition, review your storyboard yourself for terminology, consistency, simplicity, and for all other didactic and design aspects.

While you have not yet started shooting and producing the video, it's easy, fast, and inexpensive to make changes. However, if you discover mistakes only after the shooting, it will be time-consuming and costly to reshoot and edit parts of the video. Sometimes, reshooting a video may even be impossible if the presenter, the product, or the shooting location are no longer available.

Remain flexible

When shooting complex processes, it's normal that you sometimes come across things that you've forgotten to mention in the storyboard or that can be performed in a better way than you initially thought.

In cases like these, feel free to adapt the storyboard as needed.

Use the storyboard as a tool for recording the video, but don't subject yourself to it slavishly.

> **Related rules**
> *Prepare each shooting with care* 282
> *Clothing* 284
> *Lighting* 286
> *Tips for recording video and audio* 292

239

Video design

2.6.7 Add text, closed captions, or voice-over?

> In general, be very careful about adding text to a video.
>
> Unlike spoken text, which is processed in the human brain by the auditory channel, written text needs to be processed by the visual channel. However, the visual channel also needs to process the video itself at the same time. For this reason, in a video written text always distracts from the main action to some extent.
>
> Done well, however, a few short text callouts can emphasize your points for better retention of your message.
>
> In case the users of your video can't use audio, written text must replace the audio. This can make the texts quite long so that it's best to separate them from the main video to some extent, and to show them in a dedicated area as closed captions.

When you need written text

There are two basic scenarios for using written text on top of a video:

Scenario 1: Users won't be able to use audio

- No audio is available for whatever technical reasons.
- Audio isn't tolerated or would be embarrassing (for example, in an office or at a customer counter). Many people don't want to turn on their speakers in their office and maybe don't have earphones available or aren't allowed to use any at work.
- In loud places.
- Users may have trouble understanding the spoken text because their first language isn't the language of the video.

Scenario 2: Text is used to repeat or sum up what has been said for better memory

Examples:

- names of objects that the speaker is referring to
- numerical values
- warnings (see also *How to handle warnings?* 243)
- pressed keys or keyboard shortcuts
- short mnemonics

240

When closed captions are better than text on top of the video

Unlike texts within the video, closed captions:

- are shown in a dedicated place near the bottom of the video
- exactly match the spoken text

In case audio isn't available, closed captions:

- replace the spoken text
- can be helpful for users who speak the language of the video only as a second language and thus may have trouble understanding the spoken words
- can be used to improve findability and search engine optimization (when stored outside the video file)

Tips:

- If your tools support it, provide an option for users to turn closed captions on or off. Thus, closed captions are available for those users who need them, but they don't get in the way of those users who don't need them.
- If your tools don't support switching on and off closed captions, you can generate two versions of the video: one with closed captions and another one without.

When spoken text works best

The advantages of spoken text are:

- When video is combined with audio, users can process the given information through two channels: seeing and hearing. This has been proven very effective for understanding and memorizing information.
- Narration can point out certain things that can't be seen or that are often overlooked.

The disadvantages of spoken text are:

- You need a speaker (see also *Professional voice, own voice, or synthetic voice?* 251).
- Translations are time-consuming and expensive.

Tips:

- Don't describe what can be seen. Rather, complement what's *not* shown in the video. Show how, explain why.
- In case you decide to use audio, good audio quality is crucial. Poor audio ruins even the best video.

Video design

Additional tips

- Avoid cognitive overload: Synchronize shown or spoken text well with the video to relieve the users' short-term memory. Don't put any text on top of unrelated video.

  If necessary, pause the video by showing a still image until the narrator has finished or until users have had enough time to read a written text. Don't show something else as a filler that has actually nothing to do with the text.

- To make watching particularly easy, it can sometimes be better to first just display or speak the text and to show the corresponding action only when the display or speaking of the text has finished. Even though this makes the video longer, it may be well worth the extra time because it helps users focus on one thing after the other and gives them more time for mental processing.

- Make pauses. Don't speak or show too much text at a time.

- Keep in mind that not all users read as fast as you do. In addition, unlike you, users do not already know the text—and they do not know the subject either. So they need more time to think about what they are reading or hearing.

Related rules

Professional voice, own voice, or synthetic voice? 251

2.6.8 How to handle warnings?

Like with written texts, warnings must attract some special attention also in a video.

Make sure that **every user** sees or hears the warning.

Always:

- If you use audio, include the warning in the spoken text.
- Also show the warning on top of the video as text. By doing so, the warning can be seen even by users who don't use audio or who can't understand the audio for whatever reason.

    For users who *can* hear the audio, the additional text puts some extra emphasis on the message.

- Make sure that the written warning and the spoken warning:
    - appear at the same time
    - say exactly the same

General basic rules for warnings

Comply with the appropriate safety standards, depending on your product and on the country where you sell your product. These rules take precedence over the general recommendations given below. For example, if warnings follow the ANSI and ISO standards, you must add a specific safety symbol, depending on the particular hazard.

- Don't just tell users what to do or not do, but help them understand *why* they should take a particular precautionary measure. Only if users are aware of the reasons for a measure, as well as of the personal consequences and implications of not following the measure, will they take the warning seriously. For this reason, each warning must provide the following information (in this order):
    - Safety alert symbol.
    - Signal word that indicates the severity of the hazard: *Caution*, *Warning*, or *Danger* (see the following section on the types of warnings).
    - Information on what kind of danger exists, and on where the danger comes from.
    - Information on what can happen to the user, to other people, to the product, and to other things.
    - Information on how the danger can be avoided altogether, or on how at least the risk can be minimized.

243

Video design

- Always place the warning directly before the step that's dangerous or causes a hazard. If the warning comes only after a step, it may come too late. If the warning comes already at the beginning of the video, users may not remember the warning when the procedure reaches the dangerous step.
- Keep warnings short and to the point. The optimal length is one sentence. Avoid warnings that are longer than 3 sentences. Consider splitting longer warnings into 2 separate warnings.
- Address the users directly, using the imperative form. Don't use the passive voice.
- Use appropriate vocabulary to point out the possible consequences of disregarding the warning.

Types of warnings: CAUTION, WARNING, DANGER

The signal word at the beginning of the warning indicates the severity of the hazard. To make it particularly emphatic, it's often written in capital letters.

- A warning that begins with the signal word **CAUTION** indicates a hazard that, if not avoided, *might* result in *minor* or *moderate* injury. A warning with the signal word CAUTION can also refer to a situation that *could damage or destroy the product or the users' work*.
- A warning that begins with the signal word **WARNING** indicates a hazard that, if not avoided, *could* result in *death* or *serious injury*.
- A warning that begins with the signal word **DANGER** indicates a hazard that, if not avoided, *will* result in *death* or *serious injury*.

✔ Yes: ⚠ *CAUTION*
Engine may be hot.
To avoid skin burns, wear appropriate protective gloves.

⚠ *CAUTION*
Formatting the disk deletes all data.
Make a backup copy of the disk, and store the backup copy in a safe place before proceeding.

✔ Yes: ⚠ *WARNING*
Ultraviolet radiation.
May cause irreversible eye damage.
Do not look into the lamp.

✔ Yes: ⚠ *DANGER*
High voltage.
Contact will cause electric shock, burns, or death.
Disconnect the power supply before removing the panel.

2.6.9 Show the presenter?

> Showing the presenter in a video has both advantages and disadvantages. Make this decision with great care. It will have a major impact on your video.
>
> Showing the presenter can be particularly interesting if your presenter is a renowned expert in the field or if you can establish him or her as an expert. This can add a lot of credibility to your video.
>
> By using a presenter who represents a particular role, age, gender, personality, or group of people, you can convey a certain image along with your video. This is often applied in sales and demo videos.
>
> On the other hand, showing the presenter can make it hard to localize your video—especially if his or her lips can be seen while speaking.
>
> In case you decide to show the presenter, you'll also need to decide on how often and in which way to show him or her.

Advantages of showing the presenter

- A presenter puts a human face on instruction and makes the video personal.
- A well-chosen presenter can make the video very authentic.
- If you always use the same presenter(s), it can increase brand recognition.
- Having a presenter can make the content very memorable. ("I remember that bald guy putting ....")
- A presenter can make use also of non-verbal communication, such as gestures and body language. This can especially help to get users excited about a product.

Disadvantages of showing the presenter

- Translation can be a big issue. Getting a good lip-sync is extremely difficult and requires some special expertise. A poor lip-sync or simple voice-over, on the other hand, can look very unprofessional.
- The ethnic group of the presenter may cause cultural resentment.
- If the video isn't properly cut, the presenter may distract from the actual content.
- You need to find a suitable person who is willing to act in front of a camera. This is harder than to find somebody for a simple voice-over.

Video design

When to show the presenter

Focus on the content. Don't show the presenter all the time.

Best show the presenter for introductions and transitions while the focus isn't so much on the product itself but more on general things.

Tip:
To make translation easy, consider showing the presenter's lips only while he or she isn't speaking. Example: When introducing the presenter, he or she can introduce himself or herself while you show something else, then you can briefly show the presenter smiling, and then again show some other footage while the presenter continues to speak.

How and where to show the presenter

When showing the presenter, you have various options:

- You can show the presenter in a **neutral place**—for example, in front of a flip chart.

    This is rarely done because it creates quite a distance to the content of the video.

- If your product is a physical device (hardware), the presenter can stand **in front of the product**.

    Typically, this is the best option. However, it requires that the presenter is at the same location as the hardware. This isn't always the case, especially if you're using an external, professional presenter.

    Alternatively, you can also record the presenter in front of a green screen and then add the recording of the presenter on top of the video. The downsides are that you need some special equipment for the green screen and that the results often tend to look somewhat artificial.

- If your video shows software, you can record the presenter sitting **in front of the computer**—like a colleague talking to the user.

    This can typically be implemented with little effort and is very realistic, lively, and authentic.

- You can embed the movie of the presenter into a **small window within the main video** (picture in picture).

    Typically, this can be achieved quite simply if your video editing software supports it. For users, however, essentially watching two videos at a time can be very distracting. Thus, better avoid this technique.

- Instead of working with a picture in picture, better **cut back and forth between the main video and the presenter**.

    This provides a very authentic experience because it mimics the natural way people constantly move their eyes between a presenter and what he or she is showing.

Video design

Related rules

Rethink you idea of "good" 280
Male or female presenters? 250
One or two presenters? 248
Professional voice, own voice, or synthetic voice? 254

2.6.10 One or two presenters?

To make a video more entertaining, you may consider having two presenters instead of just one.

For example:

- You can have two alternating presenters.
- You can have one presenter and one commentator. If you want to make the video a bit entertaining, the second person can ask some provocative questions or add some humorous comments.
- You can create a question and answer scenario. One person acts as a "user" who asks typical questions. The other person takes the role of an expert who answers the questions and adds some additional information on his or her own if needed.
- You can combine the original narration of a presenter with comments and transitions of a professional speaker.

Combining a male with a female voice

If you're going to have two presenters, consider having a combination of a male and a female one. This way, you can combine the advantages of both genders' voices (see also *Male or female presenters?* 256).

> **Important:** If you have one "inferior" and one "superior" role (such as beginner vs. expert), carefully consider which gender will play which role. For reasons of gender equality, it may be even better to use two speakers of the same sex for this type of role assignment.

Pretending some journalistic distance

When using two presenters, you may split their roles as follows:

- One person acts as the main presenter.
- The second person takes a more skeptical role, asking provocative questions and giving critical comments.

Many users will subconsciously perceive the second person as being "neutral" even though this person in fact of course isn't.

In a sales video, you may utilize this to gain some extra credibility, to point the attention to specific aspects, or to take the mind off other things.

Video design

Related rules

Show the presenter? 245
Male or female presenters? 250
Professional voice, own voice, or synthetic voice? 254

2.6.11 Male or female presenters?

Both male and female presenters have their specific strengths and weaknesses.

In general, prefer the gender that matches the majority of users (male voice for male users, female voice for female users). This is most credible because then the presenter is most likely to be perceived as a typical user himself or herself and thus regarded as competent on the subject.

An interesting option that can add some extra appeal to your video is to have a combination of a male voice and a female voice.

Pros and cons of male voices

- A deep man's voice can have a calming effect. This can be important especially in troubleshooting situations.
- However, if the calming effect is too strong, over time a male voice can sound a bit boring.

Pros and cons of female voices

- A female voice can be clearer than a male voice even if there's some background noise in the video or in the environment where the video is watched.
- A high-pitched woman's voice can attract attention, which is particularly good for warnings and other critical information. On the other hand, it can sound a bit unpleasant over time.

Combination of a male and a female voice

A combination of a male voice and a female voice can combine the advantages of both genders' voices and add some extra appeal to your video.

For possible scenarios, see *One or two presenters?* 248.

> **Related rules**
>
> *Show the presenter?* 245
>
> *One or two presenters?* 248
>
> *Professional voice, own voice, or synthetic voice?* 251

2.6.12 Professional voice, own voice, or synthetic voice?

In general:
- A professional voice-over artist works best for rather formal videos and for sales videos.
- The voice of a subject matter expert (or your own voice) works best if the users of your videos are also subject matter experts.

For purely instructional videos, a synthetic voice can be an alternative.

Pros and cons of using a professional voice-over artist

- The voice sounds professional and is clear and easy to understand. This can be particularly important for users who speak the video's language only as a second language. Some audiences, especially for premium products, also simply expect a professional speaker.
- Professional voice-over artists often just read a script without really understanding what they are telling. Users will notice this—no matter how good the artist is. The whole video then isn't very authentic and believable.
- Your voice-over artist won't always be available. This limits your flexibility: In case you need to make changes or need to add new stuff in the future, you have no guarantee that the same voice will still be available. In the worst case, you might need to record the whole voice-over again with a different voice.
- Hiring a professional voice-over artist involves external costs. (However, note that most other persons, including you, also cost some money unless they create the video in their spare time.)

Pros and cons of using a subject matter expert's voice or your own voice

- An authentic subject matter expert may not speak perfectly but will automatically use the right articulation and inflection, which demonstrates his or her expertise to users who are also subject matter experts. A professional artist may *sound* more professional, but a normal person actually *is*.
- A not quite perfect voice can give the video (and thus the product) a more human touch.
- However, the voice must be loud and clear without any strong accent, and the person must be able to speak slowly enough. Remember that possibly not all users speak the video's language as their first language and that not

Video design

every user will be sitting in a quiet office, using high-quality speakers or earphones when watching the video.

Also, don't be afraid of using your own voice. Do you think it sounds poor? To others, it sounds just normal. It's authentic. And it's readily available.

Tip:
To make a video particularly authentic and to make the audience forgive small imperfections, consider introducing the presenter in person at the beginning of the first scene. Users will be likely to forgive "Steve from the development team" or "Anna from support" a bit of rambling more than they will forgive a nameless narrator.

Pros and cons of using a text to speech engine (TTS)

Synthetic voices may lack the human emotion needed in sales videos, but they can be well acceptable for training and documentation content.

- The voice is slow, clear, and consistent and thus easy to understand. This is particularly helpful for users who speak the video's language as a second language.
- If a video needs to be updated, you can obtain exactly the same voice again. Unlike with a recording of a human voice, the tone of the new audio snippet will perfectly match the existing audio.
- If you can install the text-to-speech engine on your computer (and only then), you can be sure that you'll have the voice available at any time.

Thus, text-to-speech is especially interesting:

- if your video isn't for marketing
- if a lot of text needs to be spoken
- if it's likely that you'll need to update the text often
- if you're on a budget

Related rules

Add text, closed captions, or voice-over? 240
Male or female presenters? 250
One or two presenters? 248
Use music? 253
Include ambient noise? 255
Tips for recording your own voice 298

2.6.13 Use music?

> Music can add some extra polish to a video and sets a specific mood.
> However, poorly selected music, or music in the wrong places, can be annoying and distracting.

Advantages of music

- Music can set a specific mood. This is often used in marketing videos. But you can also utilize this in a tutorial, for example, or you can utilize it in troubleshooting information to calm users down.
- Unobtrusive music can add that extra polish to a video that makes it appear truly professional.
- Music can fill periods of no narration and help users stay attentive.

Disadvantages of music

- Some users may strongly dislike particular music.
- If used as an introduction to a video, music wastes the users' time. This is particularly significant for short videos and can be very annoying.
- When there are spoken words, music that is playing at the same time is essentially distracting "noise." Users must then mentally separate the spoken words from the music, which can be tiring and makes concentrating on the content more difficult.

Tips for selecting the right music

- Don't choose any extravagant music but rather the usual taste—unless the video is directed at a very specific audience.
- Match the music to the message. Example: If you want to emphasize that your product is simple: use some light and easy-going music. If you want to emphasize that your product is revolutionary: use some stronger, more energetic music.
- Don't use any music that's specific to a particular culture—unless the video or its audio track are made specifically only for that market.
- Choose music whose speed matches the pace of the video.
- Use some specially designed background music. Songs with vocals, chorus, whistling, or changing volume are usually not appropriate because there are always some parts that interfere with the video.

Video design

Tips for using the music

- Use music only if it serves a particular purpose. The music must not distract from the video's content, and it must not make it harder to understand the narrator.
- No matter how great your music sounds: Don't forget that you're producing an instructional video, not a music video clip. Keep background music in the background.
- If someone is talking in the video, don't let the music get in the way. The best music in a video is the music that users don't notice even though it is there. Decrease the volume of the music while there's any narration, but don't mute the music completely in order not to get any harsh changes between periods with and without music.
- Don't be afraid to loop sections of the track. Users won't notice because the music is only in the background.
- Note that using music may require some extra royalty payments that sometimes depend on the frequency of how often the video is played.

> **Related rules**
> *Include ambient noise?* 255

2.6.14 Include ambient noise?

> Ambient noise can severely distract from the actual content of a video and can make it hard to understand the narrator.
>
> On the other hand, some ambient noise can make a video authentic.
>
> - Decrease the volume of the ambient noise while the presenter is speaking.
> - Also at other times, keep the volume of ambient noise low. Often, the best ambient sounds in a video are those sounds that users don't even notice.
> - Only include ambient sounds that have something to do with the product, its typical environment, or the actions that you perform in the video.

Examples of sounds to include

- The sounds of the device or machine shown in the video.
- The sounds made by operating the device or machine shown in the video.
- Sounds in the environment:
    - If the video is about repairing a machine: a wrench dropping on the floor.
    - If the video is about an office product: a telephone ringing in the background.
    - If the video is about a product for pets: the sounds of the corresponding animal.
    - If the video is about a product for hikers: the rustle of the wind or the twittering of birds.
    - ...

Examples of sounds not to include

- The noise from an air conditioner.
- Traffic noise through an open window.
- Talking or laughing people in the office.
- ...

Video design

Related rules
Rethink you idea of "good" 280
Use music? 253

2.6.15 Keep the video simple

> Most products are complex enough, so at least make the documentation, including videos, simple.
>
> Don't try to impress with fancy effects and with complexity. Better, impress with simplicity and clarity.
>
> Simple videos are easy to watch and save time. Users will appreciate it!

Only show what's important

Omit unnecessary detail. Don't tell everything that you might want to tell, but only tell what the users actually want to know.

- In marketing videos, this is how the product makes their lives easier.
- In instructional videos, this is how to reach a specific goal as easily and as quickly as possible.

Often, there are multiple ways of how a particular task can be accomplished. Most users, however, will only want to learn about *one* of these ways. Thus, also show only one way of doing it. Don't show any alternatives. Depending on the qualification and preferences of your audience, show either:

- the easiest way
- the fastest way
- the least error-prone way

Don't show multiple actions at the same time

Humans can focus only on one thing at a time. Thus, also show only one thing at a time.

In the rare cases when some steps really need to be performed simultaneously, show them one after the other (and communicate this clearly).

Make short, simple sentences

Don't put too much information into one sentence. Follow the rule "One idea, one sentence."

Avoid parentheses and nested sentences. Long sentences make it difficult to understand a text because they consume a lot of short-term memory.

Video design

✘ No: *The sentences in many videos, including also instructional videos, are much too long and complicated, as you can see in this example.*

✔ Yes: ***This example shows: The sentences in many videos are much too long and complicated. This includes also instructional videos.***

Be concise

Omit all words and syllables that are nothing but empty calories.

✘ No: *The cable is 10 meters in length.*
✔ Yes: ***The cable is 10 meters long.***

✘ No: *The lid has a rectangular shape.*
✔ Yes: ***The lid is rectangular.***

✘ No: *If you're a user who has experience in this field, use expert mode.*
✔ Yes: ***If you're an experienced user, use expert mode.***

✘ No: *The program isn't able to create color printouts.*
✔ Yes: ***The program can't print in color.***

✘ No: *It's necessary to enter a value.*
✘ No: *You're required to enter a value.*
✔ Yes: ***You must enter a value.***

✘ No: *You can format the table by means of the toolbar.*
✔ Yes: ***To format the table, use the toolbar.***

✘ No: *In order to print the file, choose the menu command File > Print.*
✔ Yes: ***To print the file, choose* File > Print.**

✘ No: *It takes a longer period of time to write a user manual than to read it.*
✔ Yes: ***It takes longer to write a user manual than to read it.***

Don't modify or qualify words that don't need to be modified or qualified. Unnecessary qualification doesn't add any valuable information.

✘ No: *Both sentences actually tell you exactly the very same thing.*
✔ Yes: ***Both sentences tell you the same.***

Be consistent

Always use the same term for the same thing. Using different terms can be very confusing for the users. Repeating a word in the same text is perfectly OK.

✘ No: *Click the print button. The icon only appears if a printer has been connected.*

✔ Yes: *Click the print button. This button only appears if a printer has been connected.*

✔ Yes: *Click the print icon. This icon only appears if a printer has been connected.*

If you're creating multiple videos, use the same term in each of them.

Be positive

State your points positively when you can.

- Positive sentences convey a better image of your product than negative sentences.
- Positive sentences are more engaging.
- Positive sentences are easier to understand and to remember.

✘ No: *Don't put statements in the negative form.*

✔ Yes: *Put statements in the positive form.*

✘ No: *Don't switch off the computer until you've saved your documents.*

✔ Yes: *Only switch off the computer after you've saved your documents.*

✔ Yes: *Save your documents before switching off the computer.*

Don't say "please"

When you're giving instructions, you're not asking users for a favor. For this reason, using the word *please* is inappropriate and just bloats your text. Using the word *please* can even be misleading because it implies that an action is optional, which in many cases it is not.

The only time when it's appropriate to say *please* is when you're apologizing for a problem, or when you're actually asking for a favor.

✘ No: *Please insert the card.*

✔ Yes: *Insert the card.*

Video design

✔ **Yes:** *If you get an error message, please contact support.*

✔ **Yes:** *Did you like this video? Please send your feedback to feedback@ourdomain.com.*

Use the active voice: Speak to the user

Make your videos clear and engaging. Speak *to* the users directly. Don't speak *about* them. Avoid indirect speech and the passive voice.

✘ **No:** *To create a new folder, the Explorer needs to be right-clicked. Next, users must choose "New > Folder." In the example of this video, it's named "sample01."*

✔ **Yes:** *To create a new folder, right-click the Explorer, and then choose "New > Folder." In our example, we name the new folder "sample01."*

Feel free to repeat a word

When it adds clarity, don't hesitate to use the same word over again. Do so even if it makes the sentence a bit longer. Clarity is worth the extra word.

Be clear about what you're referring to. When using words like *this*, *these*, *that*, *those*, *it*, *they*, and *them*, make sure that it's clear which subject you're talking about. If it avoids ambiguity or makes your speech simpler: Don't hesitate to repeat the subject as often as necessary.

✘ **No:** *The bomb is connected to a yellow and to a blue wire. Cut it to defuse the bomb.*

✔ **Yes:** *The bomb is connected to a yellow and to a blue wire. To defuse the bomb, cut the yellow wire.*

OR:

The bomb is connected to a yellow and to a blue wire. To defuse the bomb, cut the blue wire.

Use short, common words

Don't show off your vocabulary. Use words that are common and short.

Use uncommon and long words only if there's no simpler alternative.

For users, listening to uncommon and long words requires more mental work than listening to common words. Users who speak the video's language only as a second language may not understand uncommon words at all.

✘ **No:** *attempt*

✔ **Yes:** *try*

Video design

✖ **No:** *employ*
✔ **Yes:** *use*

✖ **No:** *indicate*
✔ **Yes:** *show, tell, say*

✖ **No:** *terminate*
✔ **Yes:** *end*

Use technical terms carefully

Technical terms speed up communication between people who share the same expertise. For others, the same technical terms are just incomprehensible.

Use technical terms only if either:

- your video is for experts
- a term is also used in other materials for the same user group as your video
- a term is used in the user interface of your product

Use strong verbs

Use strong verbs that make your texts clear, simple, and concise.

Avoid all sorts of "smothered verbs." Many people tend to use these verbs to make a text sound more "sophisticated." In your videos, do the very opposite.

✖ **No:** *The first step is the deletion of all needless words.*
✔ **Yes:** *The first step is to delete all needless words.*
✔ **Top:** *First, delete all needless words.*

✖ **No:** *You can exert influence on the printing quality by using different types of paper.*
✔ **Yes:** *You can influence the printing quality by using different types of paper.*
✔ **Top:** *The type of paper influences the printing quality.*

Related rules

Avoid visual noise 20
Use effects wisely 262

261

2.6.16 Use effects wisely

To make instructional videos more engaging, you can use a number of cinematic techniques. Almost any video editing software supports quite a few of them.

However, never forget that you're NOT making a Hollywood movie. Don't use an effect solely for its artistic appeal. Use an effect only if it supports the main goal of your video: getting the message across quickly, reliably, and with little effort for the user. Less is often more!

Don't sacrifice usability for beauty. If beauty doesn't get in the way it's perfectly OK and desirable, but it must not distract from the content.

When to use an effect

Everything that moves, heavily attracts attention. Since the Stone Age, we have been programmed to pay some special attention to sudden movements. (After all, it could be the saber-toothed tiger jumping around the corner ....)

All sorts of effects involve some form of movement, which makes them very strong. So don't overuse them. Else, they will get annoying. Remember that video is just the medium. The goal of an effective instructional video should not be to focus on the medium, but on the content.

Only use effects:

- if the effects help to focus the users' attention (and if this focus is actually needed in the particular situation)
- if the effects help to depict something more clearly

For example, it makes absolutely no sense to present software with a "cool" animated 3D effect—even though some screencasting tools support this.

If you *do* use an effect, keep the effect itself as simple as possible. Example: For the transition between two scenes, just fade to black or white, or use no animation at all. Don't add anything flashy, like Venetian blinds or a spiral, which involve a lot of movement but actually has no meaning.

Be consistent. Use only a small variety of effects and always use them in the same way. Thus, users get familiar with your video quickly and can fully concentrate on the video's content.

Overview of useful effects

Of the many effects possible, the following can make particular sense in instructional videos:

Effect	Use and application scenarios
Close-up	A close-up zooms in on a particular detail. The close-up itself is preceded by an establishing shot that shows the full scene or screen prior to the close-up. • The establishing shot is important for rooting the users' understanding of the big picture. Don't just show the close-up without the establishing shot. • The close-up itself shows as much detail as necessary.
Zoom and pan	This effect zooms in on the image and then moves the zoomed image so that the area of interest remains always visible. Tip: To provide permanent orientation for the user, from time to time zoom out to briefly show the full view.
Crossfade	A crossfade effect provides a smooth transition from one scene to another. The effect signals to the user that the scene is going to change, which can sometimes avoid confusion. However, don't overdo it. • Most video editors provide a great variety of crossfade effects. Choose a simple, unobtrusive one. • Keep the duration of the effect short. Don't waste the users' time. • Always use the same effect.
Cutaway	This effect is a brief cut to a different view, which afterward returns to the main view. You can use this technique, for example: • to show an object from a different perspective • to show a second person performing an additional action • in a software video: to show the hands of the user pressing particular keys • in a mixed software/hardware environment: to show actions performed on an operating panel or on a control station • to show someone using the product You can use a cutaway also to show some so-called b-roll for a short time (see *Tips for recording video and audio* 292). The idea is that when watching some real action, we also move away our eyes from time to time. Cutting

263

Video design

Effect	Use and application scenarios
	away to some b-roll mimics this behavior. Thus, users can "look away" right *within* your movie and won't need to turn their eyes *off* the movie.
	If a cutaway is used at the beginning or at the end of the video, it's called a *bumper*. You can use a bumper, for example:
	• to show a short common intro or outro (see *Consider creating a brand* [272])
	• at the beginning of the video: to show the narrator in order to put a face to the person speaking
	• at the end of the video: to provide a brief look behind the scenes in order to create an informal and and personal touch
Highlighting	If you want to focus the users' attention on particular elements within a video, it's helpful to use arrows, colored circles, semitransparent rectangles, or similar elements for highlighting.
	Note: In software videos, don't use the mouse pointer like a laser pointer for pointing to particular objects. Move the mouse only as a user would move it.
Highlighting the mouse pointer	On a large screen, a small mouse pointer can be hard to find. Most screen recording tools can automatically add some sort of highlighting behind or around the mouse pointer—either permanently or when clicking. These effects can significantly contribute to making the viewing of a software video less tiring.
Overlay	An overlay shows something on top of the main video.
	Overlays are good for showing things that you don't necessarily mention in the audio of the video. For example, in a software video, you can use an overlay to visualize a pressed keyboard shortcut. In a video showing hardware, you could use an overlay to display symbols, such as a lubrication symbol, a tool symbol, a torque, etc.
	Best make the overlays slightly transparent so that they don't completely hide what's below.
	Don't overuse overlays. They must not distract from the plot of the video but should only supplement it.

Video design

Effect	Use and application scenarios
Lower third	A so-called *lower third* is text located near the bottom of the screen. On TV, for example, this is often used for showing the name of a reporter. The lower third provides a consistent location for short, additional information without making things too cluttered. Tip: Best place the text onto a semitransparent rectangle, which may stretch from the left edge to the right edge or can be right-aligned. To blend it smoothly with the video, you can fade the rectangle at one side.
Time-lapse (fast motion)	This effect shows an action or process faster than it's in reality. It can visualize things that take minutes or hours in a few seconds. A time-lapse can also be used to abbreviate things that don't need to be shown in detail, or repetitive tasks that have already been shown in detail. When used wisely, time-lapses can make your video very dynamic, and they can demonstrate in a positive way that you value the users' time. Tip: Your video editing software doesn't need to explicitly support time-lapse effects. You can also add a series of still images on your timeline and then play them quickly one after the other.
Slow motion	This effect shows a fast movement more slowly. In instructional videos, slow motion is only needed rarely because actions performed by humans are normally slow enough to be visible in real time.
Greenscreen / Bluescreen	With this effect, you first record the presenter in front of some green (or blue) background. The green (or blue) color is later removed by the video editing software automatically and the remaining image combined with some other footage in the back. For example, you could use this to show the presenter in front of a machine even though the machine and the presenter physically never were in the same room. This works not only for videos that show hardware, but it works also for videos that show software. For example, here you could show the presenter in front of a big image

265

Video design

Effect	Use and application scenarios
	of the user interface as if he or she was standing in front of a big screen in a conference room.
	The downsides of greenscreen videos are that you need some special equipment and that the results sometimes tend to look somewhat artificial.
Whiteboard animation	This animation is often used in explainer videos. It typically shows some simple drawings being moved around on a monochrome surface, such as on a table or whiteboard.
	Another common effect is an animated pen that quickly draws shapes and writes text onto a whiteboard. If you use this effect, use it purposefully to spend additional time on important facts. However, don't overdo it and waste the time of your users.
	If you can be a bit informal, whiteboard animation is an efficient and effective way of creating animated explanations at little costs (use some software that's specialized in this kind of video).
	Whiteboard animation is often combined with stop motion video.
Stop motion	Stop motion scenes use paper figures, toy figures, or other small objects for illustrating something in a simple, yet entertaining way.
	Often, this is used in combination with some sort of whiteboard animation.
	Like whiteboard animation, stop motion video should only be used if you can be a bit informal.
	❶ **Important:** When using toy figures and other objects that you've not created yourself, make sure that this does not infringe any copyright.

▶ **Related rules**

Keep the video simple 257

Consider embedding also still images 267

266

2.6.17 Consider embedding also still images

Even though you're creating video this doesn't mean that the entire video needs to consist of moving images.

Sometimes, it makes perfect sense to show still images as well within a video, such as a still photo or a still drawing or infographic.

When still images within a video make sense

Good reasons for including stills into a video are:

- You simply don't have any video footage that you can use but only a still image.
- A drawing can go where a camera can't go—for example, inside a device. (The true equivalent of a drawing would be animation in a video, but this may be too expensive. A still image is typically more affordable and animation often not really needed.)
- Still images don't go by as quickly as moving images. Moving images are best to depict action. Stills are best for staying with an important point for a while.

Combining a still image with some movement

A still image in a video doesn't mean that the image has to be absolutely still.

To make the video lively, yet calm, you can pan the camera gently over the image. Or you can use a slow zoom effect on the image.

This trick is used pretty often also in documentaries on television.

2.6.18 Consider linking to further information

> Don't create your videos as isolated objects. Embed them into your global information architecture.
>
> At the end of a video, consider linking to further information and to related information. You may add:
>
> - a link to the video that you suggest for viewing next
> - a link to a help topic that details on what has been covered in the video
> - links to additional resources, such as software tools, sample data, brochures, white papers, etc.

Pros and cons of linking

Only add links where they are actually helpful.

Advantages of links are:

- Links provide context.
- Links can make it easy to find related information.
- Links can make it possible to keep videos short. Optional or related information can be given in other, linked videos.

Disadvantages of links are:

- Links attract attention and can interrupt the flow of watching the video.
- Each link requires a decision from the user: Continue or follow the link?
- If users don't follow a link, they feel that they may have missed something important. If users do follow a link, they miss the rest of the video.

Places for linking

You can add the links either:

- as hyperlinks in the video description
- below the video in the help topic that embeds the video
- in hypervideo: as hotspot overlays in the video

Video design

Related rules

Consider adding a call to action 270

Video design

2.6.19 Consider adding a call to action

> The principle of adding a call to action at the end of some piece of information originates in marketing copy and marketing videos.
>
> The idea behind a call to action can be applied to instructional videos as well. Even though you don't want to sell a product here, you *do* want to sell something else: You want to "sell" your solution. You want users to be able to use a particular feature successfully.
>
> In a video, you can implement a call to action in different forms:
>
> - You can have the performer or narrator speak the call to action.
> - You can show some text or images at the end of the video. In interactive videos, you can also add links or buttons.
> - You can combine both approaches.
>
> Tip:
> If you use hypervideo, you can even add a call to action in the middle of a video. For example, you can encourage users to perform a particular step themselves right in the middle of the video after you've explained the step. Then you can stop the video and wait for the user to finish. To continue the video, the user needs to click a button.

Phrasing a call to action

Talk to the user directly and use a strong verb.

✘ **No:** *Some additional tips can be found on our web site.*

✔ **Yes:** *Click here to learn how to save even more fuel.*

Example calls to action in marketing videos

✔ **Yes:** *Improve your technical documentation, too.*

✔ **Yes:** *Download a free trial here.*

✔ **Yes:** *Get a free quote.*

✔ **Yes:** *Buy here.*

Example calls to action in instructional videos

✔ **Yes:** *Now try it yourself: Open the lid and remove the filter.*

✔ **Yes:** *Download sample data here and create your own report.*
✔ **Yes:** *Learn next: ....*

Related rules

Consider linking to further information 268

2.6.20 Consider creating a brand

If you're going to have more than just one single video, consider branding your videos in a distinctive way.

Consider this in particular if your videos are not only published on your own web site or shipped along with your product but also made available on public video publishing platforms.

- Branding makes your videos clearly distinguishable from the videos produced by others. So it's easy to recognize the videos as the "official" source.
- If your videos are helpful, branding builds trust. Users who have had a positive experience with one of your videos will expect other videos of the same series or "brand" to be helpful as well. If there are competing videos on the same subject, users will prefer yours.
- At best, the positive experience with your videos also translates into a positive perception of your company and products as a whole.

Things that can help create a brand

You can, for example:

- always use the same presenter
- have the presenter wear some clothes with your company logo
- place something that has your company logo printed onto it in the background
- use your corporate main color (or a shade of this color) in the video—for example, for the background or for arrows or for text callouts
- consistently use the same elements and effects in all of your videos
- begin all videos with the same intro (so-called *bumper*)
- end all videos in the same way, for example, with your company logo and the URL of your web site

Don't waste the users' time

Whichever methods you use: Don't do anything at the expense of usability.

- Don't use anything that distracts from the real content of the video.
- Keep the intros and outros (bumpers) short. Keep them even shorter if:
  - your videos are part of a series and users are likely to watch several of them one after the other

- your videos are very short; users won't like to spend more time on the bumpers than on the content

2.6.21 Create small video modules

> Rather create several smaller videos than just one big one. This has multiple advantages for both the users and for you as the producer.
>
> Make each module self-contained so that it shows one complete aspect or action (but only this one).

Advantages of small video modules

Advantages for the users:

- A short video is less daunting than a long video. So it's more likely that users actually watch it.
- Many people find it difficult to stay focused on the same subject for a long time. When you split a long video into several short ones that are watched one after the other, users perceive the transitions between the videos as changes of the subject.
- The information can be viewed selectively. Users don't need to waste their time on things that they don't need.
- If indexed by an appropriate search engine, information can be accessed more quickly and directly. This mitigates one of the major weaknesses of videos compared to texts.
- Videos that are short and to the point can also be included into online help topics and replace text-based information there.
- Download times are short.

Advantages for you as the producer:

- You can combine the videos flexibly. If you create hypervideo, you can link them both in a linear way but also in other combinations. For example, you can add branches or you can link the videos in the form of a decision tree.
- You can flexibly add new information.
- You can reuse individual movie modules in various contexts without having to recreate them. The smaller the modules are, the bigger is the likelihood that a module can be reused.
- When there's a new product version, you need to update only those movie modules that are affected by the change.
- When the videos must be translated, you need to translate only those video modules that have text (both written or spoken).
- Short clips need less editing than large ones, such as cuts and transitions.

Disadvantages of small video modules

- Multiple short videos need additional file handling compared to just one long video.
- For information in a larger context, multiple modules need to be merged or linked.

Make video modules self-contained

Build video modules that are complete in a sense that they fully cover one aspect. (This aspect may be even small and part of a larger topic.)

- This makes it possible for users to watch the video module separately without the need to watch anything else.
- It makes it possible to potentially reuse the video module in various contexts.

Example: A video module shows the disassembly of a component. So this video should show exactly the complete disassembly but not more. Only then can this video be reused in the descriptions of the various maintenance procedures that all involve the same disassembly. If the video module also contained other information, this extra information might not apply in each context, and thus the video might not be reusable.

Set up a meaningful, common sample scenario

If possible, base all videos on the same sample scenario or sample data. This has a number of advantages:

- You can reuse the setting, which can save a lot of work.
- Reusing the same sample setting or data saves users mental energy. Once the users know the sample setting, they can fully focus on the content of the videos rather than on the setting.
- You can later combine the individual videos into one coherent tutorial.

Related rules

Standardize your video modules 276

275

Video design

2.6.22 Standardize your video modules

> Identify the types of video modules that you need.
>
> Standardize your video modules as much as possible according to these types.
>
> Define how each of the types should look like, which elements and effects it must include, and also what it must *not* include.
>
> Note:
> Standardization doesn't need to limit your creativity. Rather, use your creativity so that multiple video modules can benefit from it in the same way.

Advantages of standardization

- All video modules look consistent no matter who creates them and no matter where and when they are created.
- You can combine any modules and the result always looks as if everything has been created together.
- Once users have seen a few modules, they feel familiar with all other modules as well. So users can focus on the content rather than on the video itself.
- You save time: You don't need to reinvent the wheel each time.

Tip:
Create a template project and template styles according to your standard.

Example video module types

The following are some basic rules for common types of video modules. Feel free to adapt and complement both the types and the rules as needed.

Type	Rules
Teaser (Sales video)	- May show a procedure, but must not give any instructions. - Depending on the audience, may use some entertaining elements, such as doodles, paper figures, whiteboard animation, stop-motion video, and a slight dose of humor.

Type	Rules
	- May use storytelling ("This is Bob. He ....").
	- Must address customer needs and ought to focus on the benefits of the product.
	- Is segmented according to the benefits or to the unique selling propositions (USPs) of the product. Starts with the most compelling benefits to attract enough attention.
	- Is structured according to the principle "know – feel – do."
	- Ends with a call to action.
	- Not just *names* the presented benefits but also *shows* them, if possible, in order to make them more impressive and more memorable ("seeing is believing").
	- Add your own specific rules and conventions for styles and effects ....
Concept	- Must not show any procedures.
	- Must not give any instructions.
	- Includes a brief introduction that tells the users what benefits they will gain from watching the video to the end.
	- Typically includes static or animated illustrations.
	- If required, overlays the pure video recording with additional elements to illustrate the components of the product and how they work.
	- Add your own specific rules and conventions for styles and effects ....
Task	- Must not contain anything else but instructions.
	- As an introduction, uses only one sentence that names the goal of the procedure. Example: "To replace the lamp:".
	- Talks to the user directly.
	- Must use only short and simple sentences.

	- Only shows:
	  – the product
	  – the object that is manipulated by the product
	  – the tools used to manipulate the product

Video design

Type	Rules
	- Must run slowly enough so that users can see and understand every detail. - Follows the sequence of "Key – Action – Information" (KAI): 1. Indicate the action with the first frame (Key). 2. Show the steps (Action). 3. Show the result, and link to additional information if necessary. - Add your own specific rules and conventions for styles and effects ....
...	...

In addition to the specific rules, there are some things that all video module types should have in common:

- dimensions, resolution, frame rate, file format
- colors and styles of the elements added on top of the video
- narrator, music

> **Related rules**
> *Create small video modules* [274]

278

2.7 Video production

Today, you don't need any fancy video equipment to obtain very good results. If you're on a budget, even a smartphone camera can produce good video for technical documentation.

What's more important than the latest technology is the expertise on how to stage your action in the best possible way. The following tips will help you to do so:

Rethink you idea of "good" 280

Prepare each shooting with care 282

Clothing 284

Lighting 286

Tips for recording video and audio 292

Tips for presenting yourself on camera 296

Tips for recording your own voice 298

Tips for localizing and translating your videos 301

Tips for keeping your videos up to date 303

In your videos, take your time 305

Consider keeping mistakes visible 307

For information on what to put into a video and on how to best present it, see:

Video design 219

2.7.1 Rethink you idea of "good"

A video that looks highly professional and stylish is not necessarily a good video. Just look at the success of some "amateur" tutorials on popular video platforms and compare these videos' number of views and ratings to those of expensive videos made by the manufacturers of the corresponding products.

What makes these "amateur" videos successful is not the budget spent on producing the videos but:

- the fact that the videos clearly answer the users' questions without wasting their time on things that a company may want to say but actually nobody wants to hear
- a high degree of authenticity and thus trustworthiness

A "good" instructional video is not necessarily a beautiful video but a video that:

- is actually being viewed
- answers the users' real questions
- makes users more productive (even beyond answering their explicit questions)

The quality of an instructional video is determined by:

- psychological factors, especially source credibility, which is often much underestimated
- expertise on the subject
- didactics

Technical perfection and beauty are secondary.

Ways of making a video authentic

- Strictly separate instruction from marketing. If the video is a marketing video, optimizing for marketing is fine—else it is not.
- Show the product in a realistic environment rather than in a perfect studio.
- Choose a presenter who matches the audience as well as possible (age, gender, ethnic group, clothes, way of speech). Try to find a person whom the users can identify with. Select a person of average height and weight, dressed simply and with a simple hairstyle.
- Show the presenter at least briefly to build some sense of human relationship.

Video production

- As a presenter:
  - Don't be afraid to speak as you would normally speak. The more natural you act the better.
  - Don't pretend to be someone you aren't.
  - Look into the camera as if you would look at the person you're talking to.
  - Don't beat about the bush: Don't mention only the positive things but also the downsides, problems, and limitations of your product.
    Tip:
    Best combine talking about negative things with some positive advice on how to overcome the mentioned limitations.
  - Think of the users as equal and treat them as equal. Don't attempt to appear superior in any way.
  - Talk to the users directly. Use "you" or "we" ("You can ...", "We can ...") rather than third person or even passive voice ("Users can ...", "... can be done").
  - Keep everything as simple as possible. (KISS-Principle: Keep It Simple and Stupid.)

Related rules

Consider keeping mistakes visible 307

Clothing 284

Tips for presenting yourself on camera 296

Tips for recording your own voice 298

2.7.2 Prepare each shooting with care

> Unlike creating technical documentation and user assistance in written form, shooting video needs some preparation.
>
> Scout the location ahead of time. The better prepared you are, the more efficient the shooting will be.
>
> However, don't expect to be able to plan everything to the very detail. You'll always need to remain flexible to some degree.

Location and equipment

If possible, visit the shooting location in advance:

- If you know the location beforehand, you can better put together the necessary equipment (such as the lighting).
- If something is not as expected, you have enough time to adjust the storyboard prior to the shooting.

In particular, find out:

- How is the light?
- What are the room conditions like?
- Can the camera be positioned at the intended places?
- What's in the background? Is there anything that needs to be removed or covered?
- Are there any disturbances from machines or people?
- Will the shooting location be clean and tidy at the day of the shooting?

People and processes

Make sure that all machines and devices that you need to show will be fully operational on the day of the shooting:

- Will the right people be available to operate the product and for possible troubleshooting?
- If any people are to be visible in the video: Do you have their written consent?
- Is there enough fuel / an adequate power supply?
- Are there sufficient raw and auxiliary materials available?

Video production

- Is your own equipment (camera, lighting, etc.) complete and fully functional?
- Is the storyboard complete?

Product

Make sure that the product is in precisely the condition that you want to show in the video:

- Make sure that the product is available and working.
- Make sure that the product version is correct. In case the product is still a prototype, make sure that it sufficiently resembles the final version.
- Make sure that the locale of the product is correct. (Example: Correct power plug according to the target market; correct language of the user interface.)
- If the product is configurable: Make sure that you use the default configuration. With software, also check the operating system. With web applications, in addition check the browser (see *Use standard settings* 194).
- With software: Make sure that no confidential data will be visible (see *Hide private data* 202).

Performed actions

Before you begin recording the video, practice the shown actions (or have the presenter practice them).

You or the presenter don't need to be perfect. However, the fewer mistakes you or the presenter make in the first place, the less time you'll need for later editing of the video.

> **Related rules**
>
> *Create a storyboard* 235
> *Clothing* 284
> *Lighting* 286
> *Tips for recording video and audio* 292

283

2.7.3 Clothing

The presenter's clothing is an important means of giving your video the right dose of professionalism and authenticity.

The presenter should wear clothes that give him or her some authority but that are also authentic. For example, a smart guy in a fancy suit, demonstrating some maintenance work, would not be taken seriously by a target group of mechanics.

As a general rule, choose the clothes of the presenter according to the group of people he or she represents. To underline the presenter's level of authority, you may like to choose a style that's slightly more conservative than average.

Protective gear

When working with machinery, make sure that all persons visible in the video are wearing the necessary protective gear.

Also make sure that nobody is wearing any watches, jewelry, and loose clothing if this isn't permitted for safety reasons.

General dos and don'ts

- People who look good on camera stand out from the background.

    Make sure that the clothes' colors:
    - are clearly different from walls and from the product
    - at the same time also harmonize with the colors of the product and the surroundings

- Consider wearing a color that matches the main color of your brand or company logo.
- Consider wearing a T-shirt or something else with the brand logo or company logo printed on.
- Avoid stripes and check patterns because they can produce flicker.
- In case you're going to use a green screen: Don't wear anything green. In case you're going to use a blue screen: Don't wear anything blue.
- Don't wear clothing that could be considered offensive in other cultures.
- If you're wearing any clothes with text printed on: Carefully check these texts whether they might be offensive, embarrassing, or legally problematic in any way. Also check these texts for typos—many of them have.

Video production

Related rules

Rethink you idea of "good" 280

2.7.4 Lighting

> The lighting can completely change the mood or experience of a video. In instructional videos, however, you don't want any particular mood. Rather all objects and actions must be clearly visible:
>
> - The scene must be sufficiently lit but not overexposed.
> - There must not be any disturbing reflections.
> - There must not be any shadows that make it difficult to see particular details.
>
> ⓘ **Important:** When setting up the lighting, don't forget to make sure that no objects of your lighting equipment, such as cables, can later be seen in the video.

Avoid daylight

For several reasons, avoid daylight:

- It constantly changes its intensity, its color, and its direction.
- When the sun is shining, it's too hard.
- When the sun isn't shining, it's too soft.
- Sunlight only comes from one direction and thus casts shadows.
- When combined with artificial lighting, you have light with different light temperatures. This results in problems with white balancing.

Block out as much outside light as you can, and bring your own light.

If you're on a budget, clamp lights are a versatile solution:

- They can be fixed to a simple light stand and directed in a variety of ways.
- To make the light less harsh, you can put some simple diffusion material in front of the lights (such as white baking paper or a white shower curtain).
- To get some softer light, you can point the lights to a wall, to the ceiling, or to a reflector.
- Best use dimmable lamps and a dimmer so that you can adjust the intensity of the light.

If you can spend a bit more money, best use a set of purpose-built studio lights. These sets typically use large lights mounted on stands and in addition include effective diffusion screens. The lights typically provide some means of setting the intensity of the light, such as dimmers or switches to control the number of lit LEDs within each lamp.

Video production

Avoid mixed light, prefer daylight-balanced lamps

The ambient lighting mostly isn't enough.

Don't mix different light colors, such as existing room lighting, window lighting, and studio lights. This can result in improper white balance of your videos and thus in footage that looks very unnatural.

For the light temperature, typically best choose about 5000K, which almost equals daylight. If there are going to be persons visible in your videos, this will make them look quite natural. Make sure that the lamps have a high color-rendering index so that even slight color differences are rendered accurately.

Use soft (diffuse) light

Overhead lights and all other sorts of spotlights can create nasty shadows on faces and objects. So turn them off.

Use diffuse light from at least two sides. This avoids uneven illumination:

- It reduces light reflections on the subject.
- It makes cast shadows softer.

Light is described as *hard* when it's emitted more or less point-shaped.

In contrast, light is described as *soft* when it's emitted from a large surface, such as from a large studio light or from a diffuser screen.

If you don't have a set of purpose-built soft studio lights, you can build some on your own:

- You can direct a spotlight to a white surface, such as a Styrofoam (polystyrene) plate or wall, to create some indirect lighting.
- You can temporarily cover the lamp with a white shower curtain or with some white paper or other translucent material. (Tip: Best use some white baking paper, which has been designed for higher temperatures.) To attach the paper, use simple clothespins, for example.

> ⚠ **Caution:** Risk of fire. Make sure that the cover material doesn't get hot.

2-point lighting

Use at least two lights that light your subject from different sides. This makes everything visible and avoids shadows. When showing a product, using two equally bright lights often works best. If this looks too flat, you can bring in more depth by using one brighter light and one less bright one.

If needed, feel free to add additional lights from other angles.

Video production

Note:
When you show the presenter working with your product, optimize the lighting to show the *product* as well as possible rather than the presenter. Of course, the presenter should also be visible well, but when in doubt, the light on the product is more important.

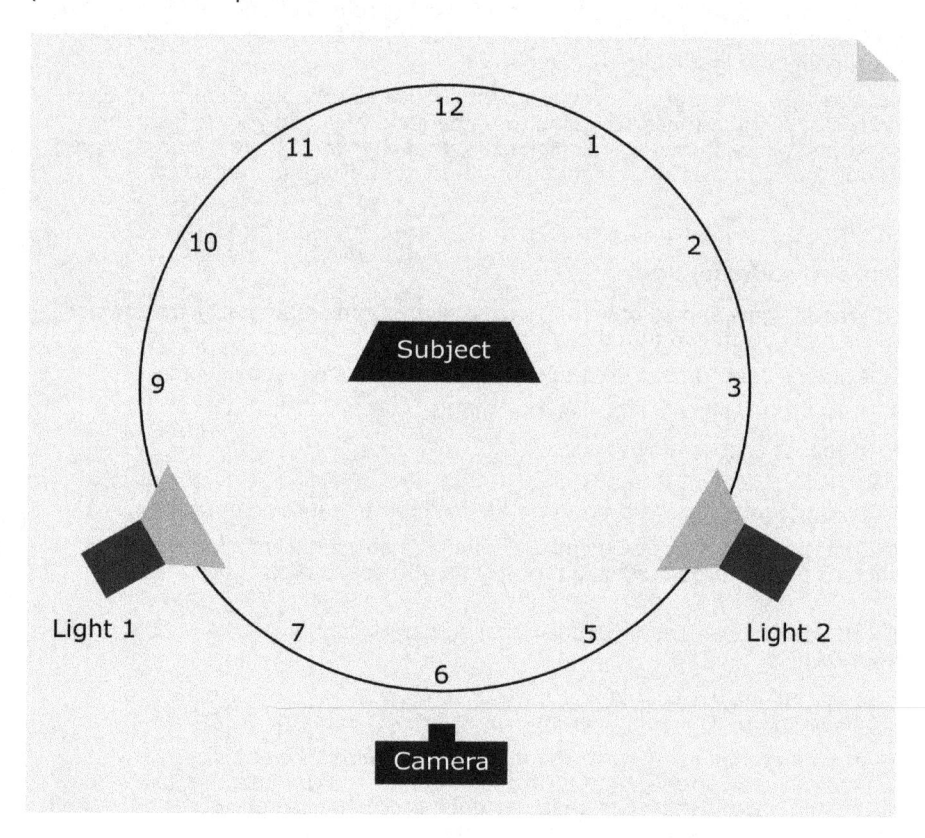

3-point lighting

When you only show the presenter but not the product, 3-point lighting can give you great, professionally-looking results. This well-known lighting method works best in interview setups, promotion videos, webinars, etc. when the subject is standing or sitting in an open room.

Video production

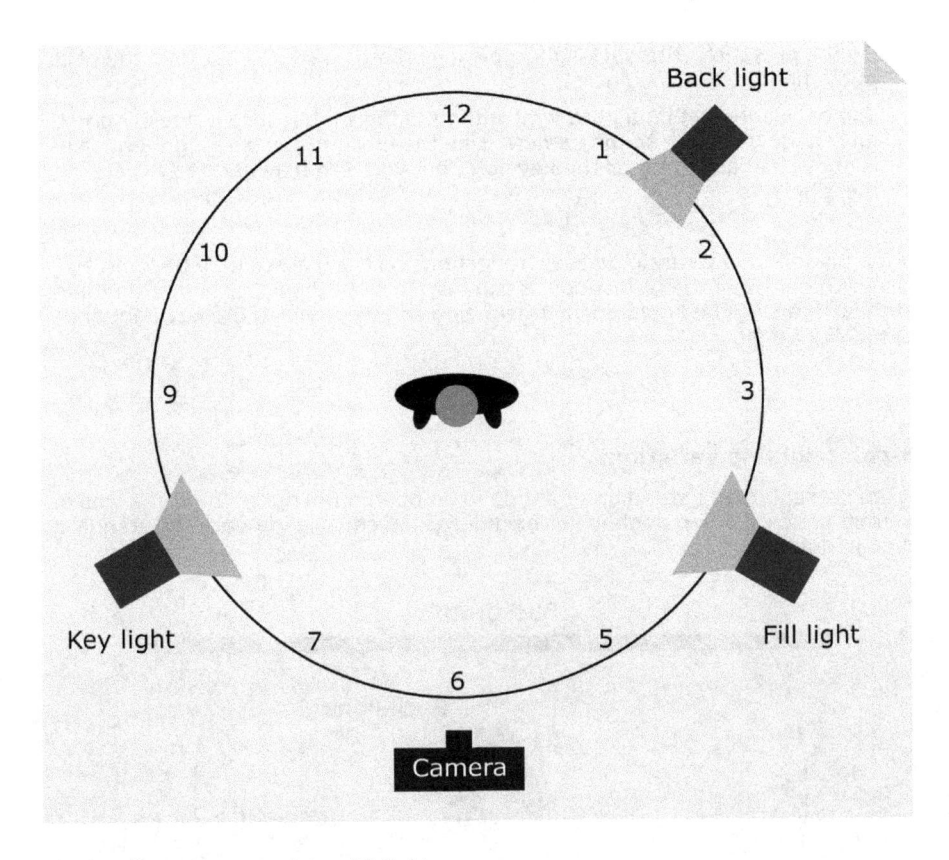

[Diagram: A clock-face circle showing positions 1–12. A subject silhouette is in the center. "Back light" is at position 1–2 (upper right). "Fill light" is at position 4–5 (lower right). "Key light" is at position 7–8 (lower left). "Camera" is at position 6 (bottom).]

The configuration consists of 3 lights:

- *Key light*: Typically, this is the brightest of all lights. It provides the bulk of light to the subject and determines the main direction of the light.

  To look natural, the key light should be placed slightly above the subject, similar to natural daylight, which in its main direction also comes from above.

- *Fill light*: Eliminates the shadows caused by the key light.

  The fill light should be strong enough to eliminate the shadows but not equally strong as the key light. The subject then doesn't look flat. For example, the nose of the presenter should still be darker on the one side, but there should be no real shadow. Typically, the fill light is about half as strong as the key light, but you may need to experiment.

  As an option, you may like to use some indirect light to make the lighting particularly soft.

- *Back light* (also called *hair light*): By lighting the subject from behind, separates the subject from the background and thus creates some more depth.

Video production

On the presenter, the back light typically creates a slight halo from the back of the head to the shoulders.

The back light can be a hard light without diffusion because it doesn't create any shadows visible to the camera. Like the fill light, the back light typically is about half as strong as the key light, but you also may need to experiment here.

Tip:
If the presenter is wearing glasses, reflections on the glasses may become an issue. To get rid of the reflections, it can help to raise the lights higher on their stands. If raising the lights doesn't help, you can try to move the key light and the fill light a bit.

3-point lighting variation

If the presenter isn't standing or sitting in an open room but in front of a wall or in front of some other single-color backdrop, you can slightly vary the standard 3-point lighting.

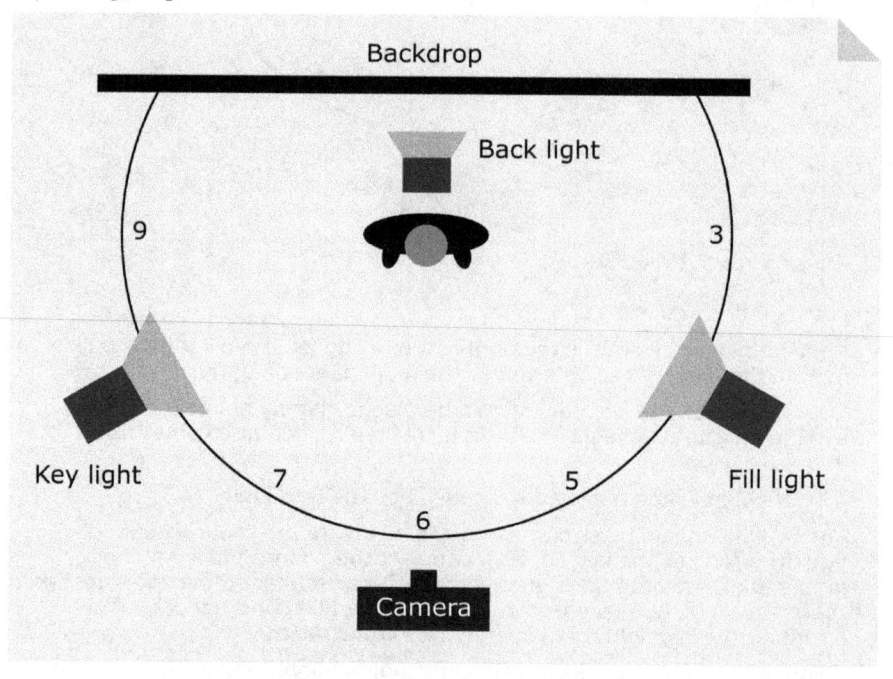

- Use a monochrome backdrop. Best, fix a roll of seamless color paper to a wall or to a backdrop stand. (Paper and stands are available from a photo supply store. Instead of paper, you can also use vinyl or polyester backdrops, which are a bit more expensive but also more durable.)

Video production

- To avoid shadows, make sure that the presenter doesn't stand immediately in front of the backdrop but with some distance. About 1.2 meters (4 feet) typically work best.
- Use two soft lights from the left and from the right that point at the presenter. If the lights are about 1 meter (3 feet) apart from each other, this is a good distance. Best place the lights slightly above the presenter's eye line. This is typically most flattering and looks natural.
- Turn off any overhead lighting. Dim the windows (if any).
- Optional: To give the lighting some extra appeal, add one more lights behind the presenter and point this light towards the backdrop. Make sure that the presenter stands in front of this lamp so that it won't be seen in the video. The lamp produces a soft halo around the presenter and eliminates any possible shadows on the backdrop.

Tip:
As an alternative to using a monochrome backdrop and a third light, you can also blur the background. You can achieve this by using a lens with a very wide aperture or by using a telephoto lens if the room is large enough so that you can place the camera far away from the subject.

Related rules

Prepare each shooting with care 282
Tips for recording video and audio 292

2.7.5 Tips for recording video and audio

> Tips on shooting videos fill books alone. However, in the case of instructional videos, you can produce good results even with a small subset of techniques.
>
> What makes shooting instructional videos comparatively simple is the fact that they need to be simple as well.
>
> Tip:
> Also look at the tips for taking good photos. Because a video essentially consists of a series of images, most of the tips for taking good photos also apply to videos (see *Tips for taking photos of technical devices* 152).

Shoot with later editing in mind

Unless your videos are very short, they will need some editing. There are a few things that can make the editing process easier and faster:

- Always shoot some buffer material at the beginning and at the end of each scene. This can later give you some extra footage if the timing requires it.
- Mark your clips to make searching for good takes easier. For example, shortly cover the lens with the hand when you've recorded a good take. When editing, you can then quickly identify that good take by the dark section.
- Even if you'll dub your videos, always record the original audio along with the video.
    - You may use the audio later to make the video more authentic (see *Include ambient noise?* 255).
    - When shooting a video together with subject matter experts (for example, a video of an assembly procedure), the audio often contains valuable hints that you later may wish to include into your own voice-over.

Shoot enough b-roll

So-called b-roll is supplemental or alternative footage that can later be intercut with the main shot. The b-roll can show something additional to the main shot and to the initial storyboard. This additional material can help to make your video less tiring because for the users it provides some "rest" from the main plot.

Also, you can use b-roll for hiding edits. For example, when a presenter makes a mistake, you need to cut the video to replace the wrong action with the correct one. To mask the cuts, you can add some b-roll to let the cuts look intentional.

Examples for b-roll that you can shoot:

- the immediate surroundings
- the presenter
- the product
- some supportive elements
- the presenter or the product from a different perspective
- a close shot or a far shot (best always record two cameras from different angles; even if your second camera is just a smartphone it can be well worth the effort)
- a person using the product
- an illustration, a flow chart, or a simplified drawing of the product
- something behind the scenes (if you want to make a video a bit humorous and entertaining, such material is often great, especially if it shows small mishaps or some other amusing things)

Also, don't forget to get close on objects and people's faces and hands from time to time. (Avoid digital zoom—it typically looks pixelated. Just get closer physically, or use optical zoom.)

Don't record your audio in an empty room

For recording your audio, look for a quiet room where no one is going to disturb you.

However, quiet rooms are often empty rooms. The problem with an empty or almost empty room is that it can produce some echo and reverb.

To resolve this, a sound dampening panel works best, but you can also use anything else that stops the sound waves from uniformly bouncing off the walls. Hang something on the walls (such as a few blankets), and put some furniture or other objects into the room. Even just some empty cardboard boxes and a carpet can make a big difference.

During the recording:

- Close the door and attach a note saying that you must not be disturbed (and explain why—or you will most certainly be disturbed).
- Mute the telephone—both the telephone on the desk and your mobile phone.

Sync externally recorded audio with a clap of hands

Simple cameras, such as smartphone cameras, often provide good video quality but poor audio quality. Thus, you often need to use an external microphone.

Video production

Getting the external audio track in perfect sync with the video doesn't require any expensive hard- or software. A basic video editor in combination with the following method will perfectly do the job. Essentially, this is the very same method that has been used in Hollywood movies for decades.

1. At the beginning of the audio recording, clap your hands twice to set a synchronization point. Wait 3 to 4 seconds, and only then start speaking. This gives you some connection material for any cuts.
2. Record the audio both with the camera *and* with the external microphone at the same time.
3. When finished, use your video editing software to import the external audio into the video.
4. Use the peaks at the synchronization point (claps) to exactly align both audio tracks.
5. Delete the audio track that was made with the camera.

Decide whether it's better to first record video or audio

If the video shows the presenter, you'll need to record both video and audio simultaneously.

If the presenter is *not* visible in the video, you can equally record both video and audio simultaneously, but you may also record them separately.

- The advantage of recording both video and audio simultaneously is that it doesn't need any postediting to get them in sync.
- The disadvantage of recording both video and audio simultaneously is that many presenters find it difficult to act and talk at the same time. This results in more imperfections or even errors both in the video and in the audio.

Thus, it largely depends on the presenter which method works better.

If you decide to record both video and audio independently (or if you have to do so because the person who performs the action and the narrator are not the same), you need to determine what to record first: the video or the audio. Each workflow has its advocates.

- With videos that show hardware, it's often better to record voice-overs first, and then to match the visuals to the pace of the audio. This is because audio is less flexible than visuals, where you can more easily slow down a section or throw in some b-roll to lengthen it.
- With videos that show software, if you're using a slide-based screen-recording tool, it's often better to first record the audio because you can freely edit the video. With full-motion-based screen recording tools, it's usually better to first record the video.

Tips for shooting stop-motion and time-lapse videos

With both stop-motion videos and time-lapse videos it's key that the camera doesn't move.

- If you have a small sandbag, fix it to the tripod that holds the camera for some extra stability.
- If you have an AC adapter for your camera, use it to make sure that the camera stays alive during the whole shooting.
- An external camera controller can make your work easier:
  - You don't need to constantly move back and forth between the objects that you move step by step and the camera.
  - It ensures that the camera exactly maintains its position when taking each picture and isn't accidentally moved when releasing the shutter.

Especially with stop-motion video, avoid daylight and even more so sunlight. A recording takes some time. During this time, the daylight and the sunlight change. This can look very amateurish in a stop motion video (light getting brighter, shadows moving along with the sum, sun hiding behind clouds, light getting darker ...).

Other than with a stop-motion video, however, in a time-lapse video this can be acceptable or even intended to visualize the passage of time.

> **Related rules**
> *Create a storyboard* 235
> *Prepare each shooting with care* 282
> *Lighting* 286

2.7.6 Tips for presenting yourself on camera

> Make an effort to appear friendly, positive, and committed.
>
> A number of things can facilitate this. You may like to complement the given tips with the things that help you personally.

How to act and where to look

Place your hands above the waistline (unless you're showing an operation with the product) and keep them away from the groin area.

While you're speaking, always look at the camera (unless you're doing an interview or demonstrating something).

How to deal with long texts

If you can't remember a text that you need to say, you can use some simple teleprompter software on a mobile phone or on a computer screen.

Another common technique is to shoot a video in bursts:

1. Speak a few sentences.
2. Look away at your script, but try to maintain your body position as well as you can.
3. Take your time to read and memorize the next sentences.
4. Continue with the next sentences, and so on.
5. Finally, cut the video with the help of your video editing software.

How to deal with mistakes

Don't throw away any good footage only because you've made a little mistake. Better just repeat the wrong action in the correct way, and later edit the video.

Tip:
When you've made a mistake, best finish the wrong sentence anyway. Then repeat the whole sentence. If you don't finish the wrong sentence, it's quite likely that you'll make the same mistake over and over again.

Tip:
When you've made a mistake, also clap your hands twice. This leaves two characteristic peaks on the sound wave in your video editor. So you can later quickly spot the places that need editing.

Video production

Related rules

▶ *Tips for recording your own voice* 298
Rethink you idea of "good" 280

2.7.7 Tips for recording your own voice

If you've decided to use your own voice for narration, make use of its main advantage: Be natural and authentic.

Being natural and authentic doesn't have to contradict a professional appearance. you're the expert! So don't be afraid to communicate your expertise to the outside world!

Preparing for the audio recording

- Use a good microphone. Poor audio quality can make a bad impression more than anything else can.
- Place a pop filter ("spit filter") in front of your microphone. This is a little shield that softens plosives or popping sounds (like T, D, and P consonants), which else may sound irritatingly sharp in your recording.
- Make sure that there's no interfering background noise, such as fans, air conditioners, traffic noise, people walking by, etc.

   Do not underestimate the importance of this point. After a few minutes, we get used to some degree of background noise and are no longer aware that it exists. However, it *is* there, and it will also be there on your recordings, where it can sound much louder and more distracting. So take your time, be quiet for a few seconds, and find out whether there's any background noise that you should eliminate.

- Set up the audio input levels carefully. Make sure that the audio doesn't "peak out," which would clip off the top of the sound wave, resulting in distortion. On the other hand, don't set the input levels too low either because then your voice would be hard to hear and would not stand out clearly enough from the background noise.
- If you need to be formal: Prepare a script. Read the text several times aloud before starting the recording.
- If you can be informal: Prepare your script only in a rough form. Don't write down every sentence in detail. This will force you to speak more freely and thus guarantees that you don't sound like reading from paper. The shorter your notes are, the more natural your voice will sound.

   Note:
   This mainly applies to procedures. When explaining a concept, it's often better to prepare a more detailed script. Explaining a complex concept can be hard to do in real time. Also, a script helps you to say what needs to be said with the fewest number of words possible.

During the audio recording

- Stand up while recording. This gives your voice an open and full sound.

  It can add some extra power to your voice to imagine speaking to a group of people rather than to an individual user.

- Maintain a constant distance between your mouth and the microphone. This is critical for achieving a consistent sound.
- Speak with a natural and clear tone.
- Don't hide your personality.
- Breathe naturally.
- Many people find it helpful to smile when they read scripts. Smiling can add more charisma and excitement to your voice.
- If you think that your natural voice is rather flat and monotone, open your mouth a bit more than usual.
- Don't rush. Small pauses are perfectly OK. Actually, many users will even appreciate small pauses because they give them some time to think about what they have just heard and seen.

  If a pause becomes too long, you can later shorten it by editing the audio track. So there's no reason to fill any pauses with disturbing "umms."

- Emphasize important words clearly.
- Last but not least: Believe in what you say (or only say what you believe in). Nothing can make your voice sound more authentic and credible.

If you've made a mistake

- Don't worry too much about small imperfections. They make your recording sound authentic and add a touch of personality. Users will remember personal messages much better than those spoken by a human robot.
- If you make a more serious mistake, don't delete the entire recording but only re-record the incorrect sentence. Best use the same technique as when having made an error in a video: Clap your hands twice to leave two visual peaks on the editing timeline, and then repeat the faulty sentence. Later, look for the peaks in your timeline and edit the audio track in your software.
- If you don't notice a mistake immediately, at least fix it as soon as possible while your tone and mood are still the same. If you wait until another day, it may be harder to reproduce the exact sound.
- If some time has passed since the original recording, match the tone of the sentence that you want to re-record by repeating the previous sentences. Repeat them several times to get in key with the tone and rhythm.

Video production

Related rules

Professional voice, own voice, or synthetic voice? 254
Tips for presenting yourself on camera 296
Rethink you idea of "good" 280

2.7.8 Tips for localizing and translating your videos

Localizing a video means fully adapting it to the specific conditions of a particular country. This includes translation into the local language. However, true localization can go even further.

If your videos will need to be localized, think ahead! There are quite a few things that you can do to make the localization and translation process faster and cheaper.

Be aware of the fact that true localization may involve more than simple translation, such as:

- replacing scenes that show local versions of the product
- replacing scenes that show a user interface in a particular language
- translating text added on top of the video
- replacing phone numbers and URLs of web sites
- replacing spoken audio
- dubbing the presenter

To facilitate localization technically:

- Store the audio and the video on separate tracks so that you can edit them independently.
- Make sure that your tools support adjusting the timing in the video according to the varying lengths of other languages.

Localizing footage and written text

When creating a video that needs to be localized, involve your localization team early even before you begin to produce the video. Have the localization team review your storyboard:

- Are your concepts going to translate well for customers in other countries?
- Are there any blunders that might get embarrassing in other countries and cultures?

When editing the video, in your video editor place all text elements on a separate track or layer so that you can later exchange them.

For software videos, best use some slide-based recording software that only takes still screenshots and then animates the mouse cursor and user input automatically. Unlike with full-motion-based recording software, you can then

Video production

exchange just the background footage without having to recreate the video completely with all of its settings in each language.

Dubbing the presenter

When planning your video, think twice whether you really need to show the presenter. Showing the presenter can make it very hard to localize your video.

- Getting a good lip-sync is extremely difficult and requires some special expertise. A poor lip-sync or a simple voice-over, however, may appear unprofessional.
- Depending on the ethnic group of the presenter, the video may impose cultural reservations. So you might need to reshoot the entire video.

If you *do* show the presenter:

- Mind the gender and the ethnic group of the presenter.
- Make sure that the presenter doesn't wear any clothing that may be considered offensive in other cultures.

Alternative: Closed captions

Instead of having a voice-over of the presenter, a cheaper alternative is to keep the original voice and to add closed captions. This solution is clearly not as user-friendly as a voice-over, however, voice actors can be expensive to hire and difficult to find, especially for rarely used languages.

You may even combine both approaches:

- Have a full voice-over for those languages that are spoken by a large number of your clients.
- Use closed captions in combination with an English-speaking presenter for those languages that are spoken only by a small number of your clients.

> **Related rules**
>
> *Tips for keeping your videos up to date* 303

2.7.9 Tips for keeping your videos up to date

> Plan and design your videos with updating in mind.
>
> Most products undergo various modifications and improvements during their life cycle. Thus, eventually you'll also need to update your videos.
>
> Software often undergoes particularly many updates in its life cycle. So keeping a software video in sync with the product can be even more of a challenge.

Things that facilitate updating videos

- Avoid showing things that will foreseeably become obsolete, such as version numbers, dates, etc. If you can't avoid that things like these are visible, cover or blur them. So users can't tell when your video was made.
- When showing a presenter, make sure that his or her clothes and hairstyle are rather neutral. So you don't run a risk that the presenter will soon look old-fashioned.
- Create short content blocks and only later combine these blocks to longer videos. So only small pieces are affected by an update.
- Save the audio and the video on separate tracks. So you can later edit them independently if only one of them needs a change.
- If updates are frequent and costs need to be low, consider using a text-to-speech engine rather than a human voice. So you can easily replace small speech fragments at any time, and you always obtain exactly the same tone and speed of the voice.
- If you're hosting your videos on a third-party platform, make sure that you can update the video without giving the updated version a new URL. So all links to the video remain valid and always point to the latest version.
- If your video shows software, you may also like to consider using simplified user interface graphics (SUIs) rather than literate ones (see also *Use a real screenshot or an illustration?* 174). In this case, it's less likely that an item that's affected by a change can be seen in the video.
- When creating a software video (screencast), don't use a pure video tool but some slide-based screencasting software. So you can more easily exchange individual screens that have changed, leaving the rest of the video unaffected.
- Always archive your video projects in their original source format—*including all used and embedded files*. So you can edit and change each video at any time in the future.
  Note:
  With many video editors, simply archiving the project isn't enough because

303

Video production

these editors only reference the actual media assets but don't embed them. So you need to archive these referenced source files as well. If you delete them, they will be missing in your video project file, and you won't be able to edit and export the final video again.

Don't be pedantic

Don't be more pedantic than your audience is. Often, it's perfectly OK to leave a video as it is if there has been only a minor change in the product.

Also, often you can keep the costs of an update low with just a small fix.

- If your video shows hardware, you can sometimes make a small fix by showing a still image that covers the original scene.
- If your video shows software, you can sometimes make a small fix by overlaying a screen with a part of a (static) screenshot that shows the updated version.

> **Related rules**
>
> *Tips for localizing and translating your videos* [301]
>
> *Standardize your video modules* [276]
>
> *Create small video modules* [274]

304

2.7.10 In your videos, take your time

Make sure that the video progresses slowly enough for the user to grasp each step. Many instructional videos fail because their speed is much too fast. Don't make this mistake.

Don't be afraid to waste the users' time. The biggest waste of time is when users need to repeat a scene because it was too fast.

(Don't confuse instructional videos with pure marketing videos. Typically, marketing videos shouldn't be longer than about 3 minutes but yet need to present quite a number of features and benefits in order to make users interested in the product. Also, a marketing video must not look boring or lengthy in any way. So a certain pace *is* needed here as long as this pace doesn't overwhelm the users.)

Timing considerations

Base the timing on the needs of the video's particular users. If possible, have somebody of the audience review your videos for their correct speed.

Find the right timing to show your actions in the video: not too slow so that users are not bored, but also not too fast so that users can easily follow the actions. Don't forget that, unlike you, users see the video for the first time and are new to the subject of the video.

- If you demonstrate how to perform an action, make it slow enough for the users' eyes to follow. If you're creating a software video (screencast), don't move the mouse pointer hectically from one place to another.
- If you add text on top of the video or if you add closed captions, give users enough time to read the text.
- If you add audio, pause the actions in the video until you've finished speaking.
- Add additional small pauses for mental processing—especially after explaining complex correlations or complex steps.
- If a significant proportion of users don't speak the video's language as their first language, give them even more time.

Tip:
To find the right pace when recording the video, speak the text aloud (even if you're going to add the final audio later). The closer you come to getting the timing right from the beginning, the easier editing will be later.

Video production

Player buttons for pace control

If the video authoring tools and the video player that you use support it, consider adding cue points and video player controls for pace control:

- a button to pause a scene
- a button to repeat a scene
- a button to interrupt a scene and to jump to the next one

When creating interactive videos (hypervideos), provide built-in stop points so that the users can perform a shown action themselves before the video continues.

Don't start anything automatically

Don't set a video to play automatically as soon as the page that embeds the video has loaded. In particular, don't do so if the video contains audio.

- Users are sure to miss the beginning. This leaves no good feeling and creates knowledge gaps—or users need to restart the video, which is also frustrating.
- For users, it can be very annoying and embarrassing when in a quiet office, suddenly some loud music begins to play or when all of a sudden a presenter begins to talk. Do you want a user to blush with shame? This is a safe method!
- The movement of the images in the video immediately attracts the attention of the users. While the video is playing, it's almost impossible to read any text nearby.

> **Related rules**
> *Keep videos short* 224

306

2.7.11 Consider keeping mistakes visible

Have you made a mistake while demonstrating an action? Do you thus need to repeat the shot? Wait a minute ....

If users are likely to make the same mistake, consider showing your mistake in the video (and make it clear that it was a mistake).

By showing a typical mistake:

- users will remember the pitfall much better than by just mentioning it
- users can also learn how to correct the mistake

You may even use mistakes on purpose and add the most frequent ones to your storyboard right from the beginning.

General rules of when to show mistakes—and when not to show them

- Be careful with showing mistakes in purely instructional videos. These videos need to be short and to the point, and the shown steps are often immediately imitated by the users. Thus, any mistake shown here may also be immediately imitated.
- Good places for showing mistakes, however, are general tutorials.

Extra tips for dealing with and for showing mistakes

- Make sure that it's clear that what was shown was a mistake. Best, point out the mistake even before it's shown so that users don't imitate the wrong action themselves.
- Don't elaborate on the mistake more than necessary. Keep the scene short. The focus should clearly be on how to perform the action correctly, not on how to perform it incorrectly.
- In a rather informal video, mistakes can be good places for adding a sense of humor to your video.

Related rules

Rethink you idea of "good" 280

2.8 Interactive content

In many cases, online documentation is essentially only a manual on screen. But Online documentation can do so much more. It's running on a computer! It's software! It can do anything that a computer can do!

Stop thinking in terms of traditional user manuals. Bring your content to life and make it interactive.

However, one word of caution:

As the word says, interaction also means *action*. To take action, users must be motivated. To be motivated, there must be some clear benefit. Therefore, only make things interactive if this actually provides some clear extra value. Don't provide interactivity just for its own sake.

You have the following options:

- *Interactive 2D images* [310] are clickable or hoverable images in general—both drawings and photos. The xy position of the mouse pointer is evaluated and triggers an action. Most often, interactions display additional information in one way or another.
- *Interactive 3D images* [313] show a 3D model of an object and its components. The model can be manipulated in all 3 dimensions. For example, users can rotate the object, isolate its components, and often also run 3D animations.
- *Interactive video (hypervideo)* [316] contains embedded, interactive anchors. By clicking these anchors, users can navigate within the video or trigger other functions.
- *Augmented reality* [318] provides a composite view by superimposing an image or other data on a user's live view of the real world.
- *Scripts and more ...* [322] can perform any actions for users, such as calculate values or look them up in a database, interact with the product, or guide the users through an e-learning course.

2.8.1 Interactive 2D images

Interactive 2D images are clickable or hoverable. The xy position of the mouse pointer is evaluated (hence "2D"). The image itself, however, may show either two-dimensional objects or tree-dimensional objects, and it may either be a drawing or a photo.

Depending on where the click is made, it can trigger specific actions. This enables progressive disclose of additional information and exploration of objects by the user.

For users, this means that the benefits of the interactive image come at a cost of some extra work and investment of time. So it's crucial that the interactive image provides (and communicates) some obvious benefit. Else, users will only consume it as a static image.

> **ⓘ Important:** From looking at the image, it's not necessarily obvious that the image is interactive. Thus, provide some visual or text-based clue that users can interact with the image.

Possible functions and examples

Think of your image as a piece of software, not as a static image printed on paper.

For example, you can:

- Progressively disclose more information: Clicking or hovering over certain parts of the image can show additional information. (Example: Hovering over parts shows the parts' names and order numbers.)
- Progressively remove parts by clicking them in the image.
- Filter the image for particular content.
- Change the perspective.
- Run an animation.
- Change sample data or modify a state.
- Switch between a photo and a drawing (which can be simplified compared to the photo).
- Switch between two photos or between two drawings either to show additional components or to remove certain components. (Example: The initial image shows a machine in its normal state. A second image shows the same machine from the same perspective but with its housing removed.)
- On a chart: Dynamically show auxiliary lines at the mouse pointer to facilitate the reading of the values at the axes.

- Show a dynamic magnifier to be able to view certain parts of the image in more detail.

Interactive images can even simulate a complete user interface. Example: An image shows the controls of a photo camera plus a photo. Users can click the image of the camera to change the camera settings. The results can then immediately be seen in the photo. So users don't need to read text but can explore the product without any risk of making errors. Thus, they are likely to discover functions that else they would not find.

Another application is the use of a clickable image as a graphical table of contents. An interactive image can provide directed access to particular information without having to name the components in the image. Omitting these technical terms can make the whole document much simpler and avoids mistakes. All that users need to do is to click the corresponding component in the image.

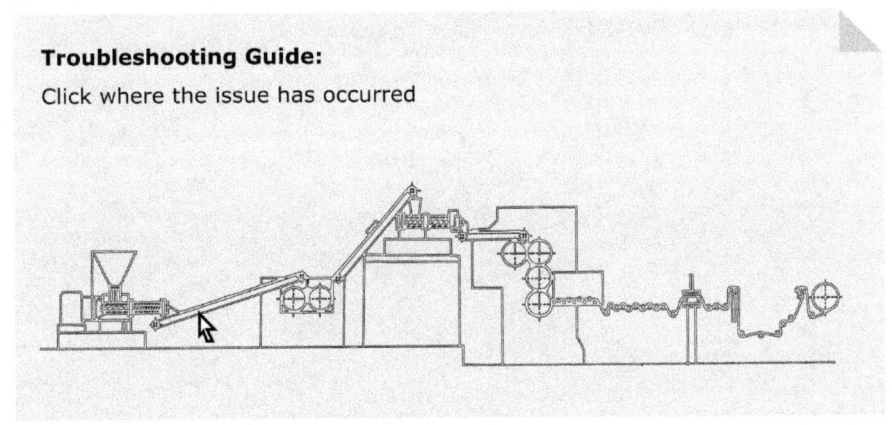

Troubleshooting Guide:

Click where the issue has occurred

Advantages and disadvantages

Compared to non-interactive images, interactive ones:

- inspire exploration
- can contain more information in additional layers
- require users to take action
- require some time for exploration
- may go unnoticed, so users may see only the static version of the image
- are more complicated and thus more expensive to create
- don't work on paper, so if you want to provide the same information also in a printed form, you need a replacement strategy

  You may either:

311

Interactive content

- omit the image altogether, losing its information—usually not recommended because if the image is that unimportant you shouldn't invest into making it interactive in the first place
- show the most simple version of the image plus some text—recommended if the amount of interactive information is large
- show the most complete version of the image—recommended if there's only little information added to the image interactively

Related rules

Images in general 47
Images of hardware 135
Interactive 3D images 313

2.8.2 Interactive 3D images

Interactive 3D images show a 3D model of an object and its components. The model can be manipulated in all 3 dimensions.

Users can zoom and rotate the shown object, and they can apply various functions, such as isolate individual assemblies and components or run an animation.

Unlike with static images, users are not limited to the perspective provided by the author but can explore the object in various ways. This even works for animations. Other than with a video, when users interrupt a 3D animation, they can change the position of the camera, zoom in or out, and then continue or rerun the animation from the new perspective.

If you have existing CAD data from development, you can typically create interactive 3D images at very reasonable costs.

If you use 3D PDF as the file format, you can embed your 3D images into normal text documents. Users can then use a free PDF viewer to view the document along with the interactive 3D images.

The key *strength* of an interactive image is also its main *weakness*: For users, viewing an interactive image means having to take action. So it's crucial that an interactive 3D image provides and communicates some obvious benefit.

> **ⓘ Important:** From looking at the image, it's not necessarily obvious that the image is interactive. Thus, provide some visual or text-based clue that tells users that they can click and manipulate the image.

Possible functions

Depending on the used file format and image viewer, users can, for example:

- zoom
- rotate
- change the perspective
- change the lighting on the scene
- run an animation (examples: step-by-step disassembly of a component; device in operation with its moving parts inside)
- filter components (and hide all others)
- highlight components
- isolate assemblies and components

Interactive content

- measure components interactively
- display metadata, such as part numbers or order numbers
- display a parts list
- synchronize the parts list with the image so that when a part is clicked in the image, it's also highlighted in the parts list and vice versa

To make an interactive 3D image particularly user-friendly, you can preconfigure specific views and make them accessible with a click of a button. Also, you can add some buttons to start preconfigured animations.

In 3D PDF, you can even change views and run animations in the *image* from links within the document's *body text*.

Example:

To view the connector, turn the object to the left.

Advantages and disadvantages

Compared to non-interactive images, interactive 3D images:

- can contain a lot more information and provide various additional features
- are ideal for exploring spatial relationships
- are more complicated to create
- are more complicated to use; so some users may have trouble using all functions of the image viewer
- require users to take action, so users need some significant motivation to explore the image in detail
- may go unnoticed, so some users may see only the static version of the image
- don't work on paper, so if you want to provide the same information also on paper, you need a replacement strategy; you may either:
  - omit the image altogether, losing its information—usually not recommended because if the image is that unimportant you shouldn't invest into making it interactive in the first place

Interactive content

- provide only one static frame of the image—recommended if it's optional for users to see the object from different perspectives
- provide a series of static frames of the image—recommended if it's necessary for users to see the shown object from different perspectives

Related rules

Images in general 47

Images of hardware 135

Interactive 2D images 310

2.8.3 Interactive video (hypervideo)

> Hypervideo (hyperlinked video) is video that contains embedded, clickable link anchors. Hypervideo combines linear video with a non-linear information structure. It enables users to make choices and to trigger actions based on the video's content and on the users' needs.
>
> Users can thus:
>
> - take their own paths through the video, including branches, omissions, and loops
> - access additional information directly from within the video
>
> To make use of a hypervideo, users need to interact with the video, which some users might not do. For this reason:
>
> - Only use hypervideo if it provides some clear extra value compared to normal, linear video.
> - Don't expect all users to interact with the video. Also provide an automatic mode for those users who don't actively participate. Make sure that this automatic mode covers at least the most important points of the video.

Possible interactions

In principle, you can do the same things in a hypervideo that you can do in the hypertext on a web site or in a help system,.

In particular, you can:

- Stop the video automatically at a certain point and wait for the user to take action. For example, you can stop after each step of a procedure and wait until users confirm that they have performed this step. Or you can stop after a warning and wait for users to confirm that they have read and understood the warning.
- Add Buttons or links that jump to another position on the timeline of the video, thus skipping or repeating content.
- Display a navigation menu.
- Display a search form.
- Start another video.
- Open a page within the documentation.
- Open an external document, such as a PDF file.
- Link to a web site.

- Pass information to a web site, for example, the order numbers of needed spare parts.
- Let users answer questions (quiz).
- Trigger functions on the product (if the product is software or software-driven).
- Influence the HTML page that embeds the video, for example, show or hide particular paragraphs, synchronize a navigation menu, etc.

Specific advantages and disadvantages

In general, the advantages and disadvantages of hypervideo are the same as with standard, linear video (see *Which medium to use?* 13). However:

- Users can find particular information faster and view it more selectively.
- Users can more easily skip irrelevant content.
- Users need to take action, which not all users may be willing to do.
- Hypervideos are more difficult to create than standard, linear videos.

> **Related rules**
> *Video design* 219

Interactive content

2.8.4 Augmented reality

Augmented reality (AR) superimposes an image or other data on a user's live view of the real world, thus providing a composite view.

This can be achieved either:

- **With the help of some special glasses:** The user sees the real object in combination with the augmented information while having both hands free.
- **With the help of a tablet computer or smartphone:** The user points the camera at the object and sees the object in combination with the augmented information on the display. This approach doesn't require any special eyewear, but the user needs to hold the tablet or smartphone with at least one hand.

AR can both *identify* real-world objects and *point to* real-world objects.

> ⓘ **Important:** The presence of AR capabilities within user assistance isn't evident. So you need to add some hints that AR is available, and you need to explain how to use it.

Use of AR for embedding information

You can use augmented reality to provide context-sensitive information on hardware components. When a user looks at an object, the AR application can recognize a component and can then automatically display:

- information about what the component does
- information about how the component can be used
- technical data
- links to access additional information
- translations of the texts that are printed onto the component

If an AR application is connected to the system, the AR application can even show current process data. This can, for example, be helpful for troubleshooting.

Interactive content

Tire - front right

Current pressure:
200 kPa (= 30 PSI)

Recommended pressure:
230 kPa (= 34 PSI)

Videos:

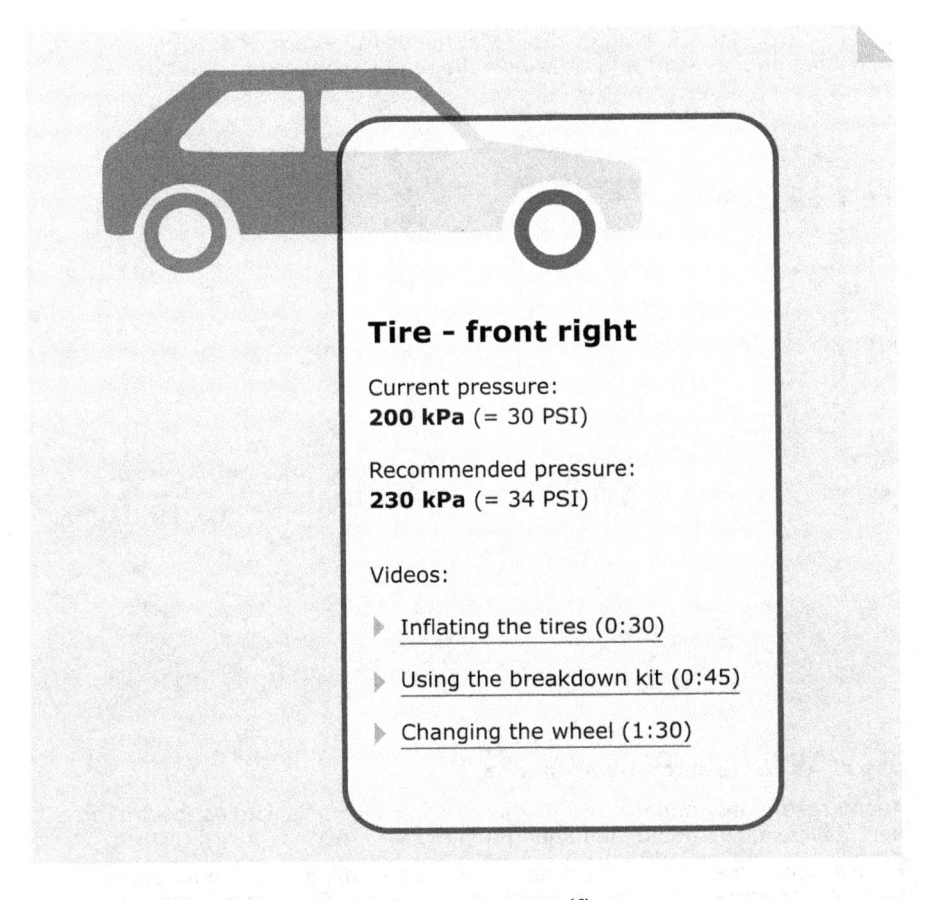

▶ Inflating the tires (0:30)

▶ Using the breakdown kit (0:45)

▶ Changing the wheel (1:30)

The given information can be highly context-specific:

- The AR application recognizes the exact product variant and particular component that the user is looking at. The user doesn't need to manually search the documentation for the applicable information.
- If the AR application works with a user login, it also knows the role of the user. Thus, for example, if the user is an operator, the AR application can show operation instructions. If the user is a technician, the AR application can show service instructions.

Use of AR for identification

You can use augmented reality to spot a particular object. For example, within some maintenance instruction, users could point the camera of their mobile device to a machine. The AR application then superimposes an image that high-

Interactive content

lights a particular nut, bolt, or cable that needs to be removed. Even a video or an animation may start and show how the removal needs to be performed.

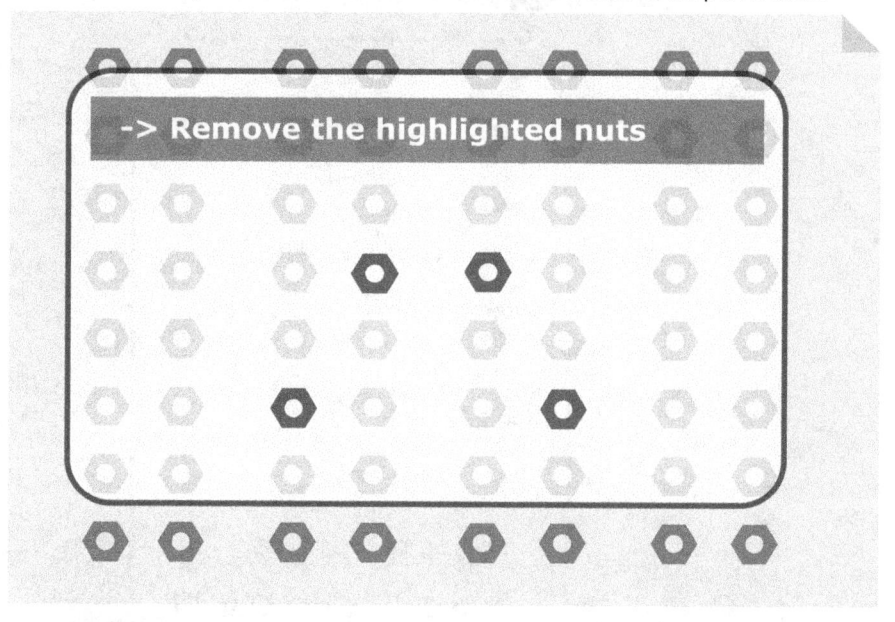

Use of AR for remote maintenance

Augmented reality applications enable users or technicians to connect to an expert via live video, audio, and annotation tools.

For example, when a technician has trouble performing a particular operation, he can connect remotely to a subject matter expert. The expert gets a live view through the technician's glasses or camera. The expert can talk to the technician, can point to elements that are then augmented to the technician, or even animate elements to exactly demonstrate what to do and how to do it.

Use of AR for complementing printed documents

You can use augmented reality even to complement printed documentation. Users can then scan an object in a printed document to retrieve additional information.

For example, when users buy a piece of furniture, such as a table, it can come with only very brief printed assembly instructions. Users who need more guidance can use AR for additional assistance. When a user scans an image in the printed assembly instructions with a smartphone or tablet, AR can run a short video or animation of the corresponding procedure, or AR can display an interactive 3D model. Also, you can provide some extra tips for usage and care, or you can link to accessories and related products.

Advantages of AR

- Augmented reality applications can automatically identify real-world objects and thus connect the real world to the documentation and vice versa.
- Augmented reality applications are highly context sensitive. For users, this saves time and avoids errors by correctly identifying objects.
- Augmented reality applications tightly integrate information with the product. While performing an action, users don't need to look away into a manual. User assistance is immediately available.

Disadvantages of AR

- Augmented reality requires some special technology and 3D models of the covered objects.
- Augmented reality requires some special hardware (glasses or tablet or smartphone).
- Augmented reality is significantly more expensive to develop than static content.
- Users need to learn how to use the particular AR application.
- To some degree, the augmented information may distract from work and thus cause extra hazards.

Interactive content

2.8.5 Scripts and more ...

> In online documentation, in addition to interactive images and video, you can implement even much more, Don't think of "writing a manual." Think of "programming user assistance."
>
> You can implement anything that a computer can do. There are hardly any limits. Get creative!
>
> The following are just a few ideas for your inspiration.

Calculators and assistants

Instead of providing static content and letting users sort out what's relevant, you can use some program logic to do the work for them.

A simple form can sometimes replace a long table or a whole series of charts by calculating the values using a formula or by querying a database.

✖ **No:**

To find out C, look at the corresponding chart or table:

322

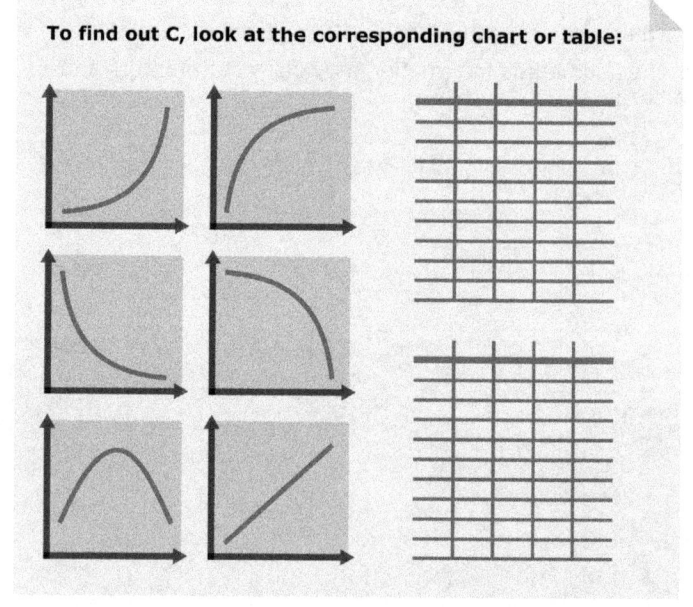

✔ **Yes:**

> **To find out C, enter A and B:**
>
> Value of A: `1001`
> Value of B: `99.99`   →   **C = 290**

A form or a series of questions can help with troubleshooting rather than having users browse hundreds of error descriptions themselves.

> ## Troubleshooting assistant
>
> To quickly find a solution, please answer the following questions:
>
> What color is the control light on the front panel?
>
> ○ Green
> ◉ Red
> ○ Dark (off)
>
> Is there a beeping sound when you turn on power?
>
> ○ No
> ◉ Yes: two beeps, a pause, two beeps, ...
> ○ Yes: multiple similar beeps
>
> **Query support database**

Experiments and sandboxes

Instead of describing something with abstract words, you can let users experiment to give them a better feeling of how things work together.

Interactive content

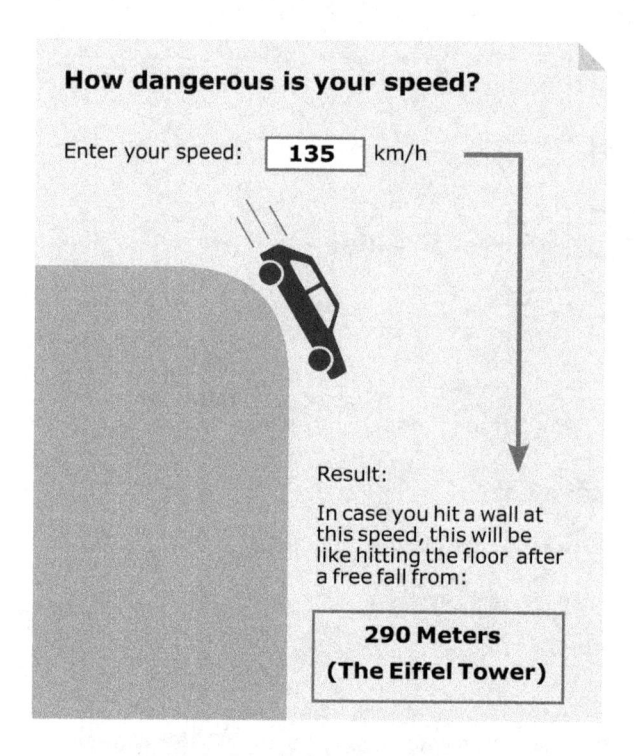

How dangerous is your speed?

Enter your speed: `135` km/h

Result:

In case you hit a wall at this speed, this will be like hitting the floor after a free fall from:

290 Meters

(The Eiffel Tower)

When describing an application programming interface (API), you can embed a simple interface for testing commands.

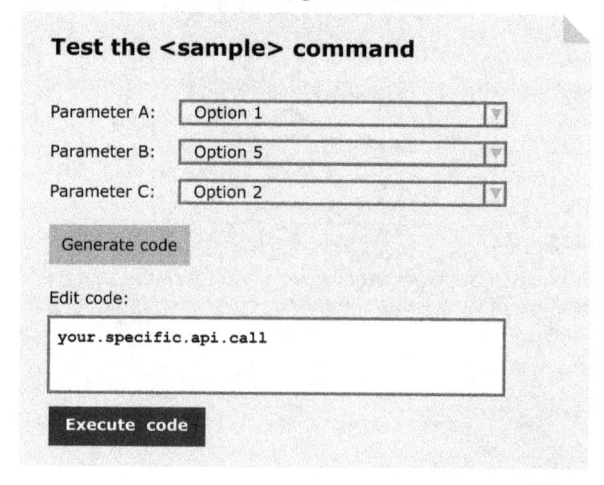

Test the <sample> command

Parameter A: [Option 1 ▼]
Parameter B: [Option 5 ▼]
Parameter C: [Option 2 ▼]

[Generate code]

Edit code:

```
your.specific.api.call
```

[Execute code]

Micro e-learning

To support active learning, e-learning uses interactive components, such as exercises and quizzes.

Other than technical documentation, traditional e-learning or CBT (computer-based training) is typically used prior to working with the product but not for quick troubleshooting and answering very specific questions. However, on a small scale, sometimes the idea of e-learning also makes sense in technical documentation:

- in parts of the documentation that aren't read during work, such as in tutorials and in getting started guides
- when what's taught will be needed many times so that it can be considered "the basic knowledge" for the use of the product
- when you can assume that users have a strong motivation for learning rather than for just looking things up when needed

E-learning elements that can be useful within technical documentation are:

- specific interactive exercises
- quizzes where users can practice and test what they have learned

You can either embed these elements into a help topic, or you can embed them into a hypervideo.

Microgames

Gamification can add elements to documentation and blends learning with entertainment.

You can embed little games into help topics that let users practice what they have learned. You can even add some social components here: For example, you can set up a high score list where users or user groups (such as the co-workers within the same company) can compete with each other.

However, microgames only make sense:

- if this fits the image that your product is supposed to convey
- in sections of the documentation that are used for learning rather than for looking things up quickly
- if you can design a game that is both fun *and* instructive

Be particularly careful if your product is used for business. Some employers may not like the idea of their employees playing games even if this helps them with their jobs.

Thus, use cases for microgames are quite rare. Yet there *are* some special occasions where they can be very effective.

3 References and further information

This is the end of the book—but it's not the end of the story.

I hope that this book could make you aware of the key factors that account for good user assistance, and I hope that the book will become your regular companion when creating, reviewing, or revising your own content.

As this book is about best practices, it's in the nature of things that I can't quote any single publication or study for a particular recommendation. Rather, the gist of the book is based on the overlap of what can be seen as today's state of the art. Inspired by hundreds of conference presentations, articles, blog posts, social media contributions, personal and forum discussions, as well as my own 25+ years of professional experience in the field of technical communication.

Many thanks to all who shared their expertise. I am sorry that I cannot list each of these sources individually. It would simply go beyond the scope at this point.

If you want to dive in deeper yourself and are looking for the best web sites, forums, conferences, and professional organizations in the field of technical communication and user assistance, you can find an up-to-date, free tool and web guide on my web site at *www.indoition.com*.

You can also find there comprehensive lists of the best authoring tools as well as links to other software for technical communicators.

In case you need any individual advice, or if you have any comments or suggestions for future versions of this book, please don't hesitate to contact me in person at: Marc.Achtelig@indoition.com

Thank you.

Yours Marc Achtelig

Index

A

ambient noise
  in videos 255
animated images
  file format 119
aperture
  impact 157
audio
  ambient noise 255
  music 253
  recording tips in general 292
  tips for recording own voice 298
  voice 251
augmented reality
  use 318
authenticity
  of videos 280

B

backdrop
  for photos and videos 155
  for videos with a presenter 290
black and white
  vs. color in images 107
borders
  designing image borders 62
  when to add around an image 62
brand
  creating for videos 272
b-roll
  shooting 292
  using 292

C

call to action
  adding to videos 270
callouts
  formatting 85
  pros and cons 82
closed captions
  use in videos 240
close-up
  use in videos 262
clothing
  of the presenter 284
color
  optimizing for printing on black and white printers 109
  using in images 25
  vs. black and white in images 107
color correction
  applying on photos 164
color mode
  of images 116
confidential data
  hiding in screenshots 202
consistency
  of image sizes 65
context
  indicating in an image 67
controls
  needed in video player 228
converging verticals
  avoiding 159
crop and pan
  use in videos 262
cropping
  photos 165
  use in videos 262
crossfade

329

crossfade
  use in videos 262
cutaway
  use in videos 262
cutaway view
  using 143
CYMK
  color mode 116

D

data
  hiding private in screenshots 202
  showing in screenshots 199
dead space
  getting rid of in screenshots 189
diagrams
  file format 119
dimension lines
  formatting 93
drawings
  tips for creating 128
  vs. photos 136
  ways of keeping simple 22
dynamics
  indicating in images 146

E

effects
  in videos 262
e-learning
  use in technical documentation 325
emphasizing
  items in images 77
experiments
  use in technical documentation 322
exploded view
  using 144

exposure triangle
  with photos 157

F

fast motion
  use in videos 262
figure numbers
  formatting and building 105
  necessity 104
figure titles
  formatting 103
  necessity 101
  phrasing 102
file format
  choosing for images 118
  for animated images 119
  for diagrams 119
  for icons 119
  for infographics 119
  for photos 118
  for screenshots 118
  for symbols 119
flashing light
  indicating in images 148
formats
  for callouts 85
  for dimension lines 93
  for legends 85
  for text callouts 85
forms
  use in technical documentation 322

G

games
  use in technical documentation 325
graphic artists
  tips for working with 131
graphics

graphics
 tips for creating 128
gray card
 using 162

H

hardware
 images of 135
highlighting
 use in videos 262
hypervideo
 use in technical documentation 316

I

icons
 designing 31
 file format 119
 finding 29
 pros and cons 28
 using 28
images
 borders 62
 callouts 82
 color mode 116
 color vs. black and white 107
 emphasizing important items 77
 facilitating translation 125
 facilitating updates 121
 file format 118
 guiding the viewers' eyes 70
 in videos as poster image 233
 in videos as sill images 267
 indicating dynamics 146
 indicating movement 146
 indicating size and context of the subject 67
 integrating into page layout 55
 legends 82
 number to include 50
 of hardware 135
 of software 167
 page breaks 58
 perspective 138
 photos vs. drawings 136
 position 51
 pros and cons 14
 purpose 49
 repeating 51
 resolution 111
 reusing (single source publishing) 123
 showing the interior of objects 143
 size in general 64
 text formatting 95
 text including 33
 text writing 37
 thumbnails using in online documentation 65
 tips for creating 128
 ways of keeping simple 21
 zooming in on details 79
infographics
 file format 119
interactive content
 use in technical documentation 309
interactive images
 2D 310
 3D 313
interactive video
 use in technical documentation 316
interior
 showing of objects 143
ISO
 impact 157
item numbers
 formatting 85

K

keystone effect
 avoiding 159

331

L

labeling
  callouts vs. legend 82
layout
  visual weight of images 60
legal pitfalls
  with images and videos 44
legends
  formatting 85
  pros and cons 82
length
  of video modules 274
  of videos 224
light
  indicating in images 148
lighting
  for photos 152
  for videos 286
links
  in videos 268
localization
  cultural differences with visuals 41
  of images in general 125
  of screenshots 214
  of videos 301
lower third
  use in videos 262

M

medium
  which to use 13
micro e-learning
  use in technical documentation 325
microgames
  use in technical documentation 325

mistakes
  handling when shooting a video 296
  showing in videos 307
mouse pointer
  when to show in a screenshot 183
movement
  indicating in images 146
music
  in videos 253

N

navigation
  in videos 226
noise
  in videos 255
numbers
  figure numbers 104

O

overlay
  use in videos 262

P

page breaks
  with images 58
pagination
  with images 58
people
  showing 40
persons
  showing 40
perspective
  for drawings 138
  for photos 152
phantom view
  using 143

photographers
  tips for working with 131
photos
  backdrop 155
  color correction 164
  cropping 165
  exposure triangle 157
  file format 118
  gray card using 162
  ISO - aperture - shutter speed 157
  keystone effect 159
  lighting 152
  resizing 164
  session preparing 150
  setup for shooting photos of large machinery 157
  tips for editing 164
  tips for taking 152
  trimming 165
  vs. drawings 136
  ways of keeping simple 22
  white balance 162
player controls
  in videos to provide 228
poster image
  for videos 233
presenter
  clothing 284
  male vs. female 250
  showing in video yes vs. no 245
  single vs. two 248
  tips for presenting on camera 296
private data
  hiding in screenshots 202
procedures
  as an image 35

R

resizing
  photos 164
resolution
  of images 111

RGB
  color mode 116

S

sandboxes
  use in technical documentation 322
screenshots
  avoiding confusion with the UI 193
  data hiding 202
  data showing 199
  faking 205
  file format 118
  general recommendations 169
  getting rid of dead space 189
  hiding the browser window in 197
  mouse pointer 183
  number to include 169
  optimizing for particular purpose 207
  pros and cons 169
  tips for taking 209
  translating 214
  use cases 170
  using 167
  using standard settings 194
  vs. illustrations 174
  ways of keeping simple 23
  what to show 177
  window size 186
script
  design tips 238
  need for 235
scripts
  use in technical documentation 322
sectional drawing
  using 143
setup
  for shooting photos of large machinery 157
  lighting for photos 152

333

setup
  lighting for videos 286
shooting
  preparing photo shooting 150
  preparing video shooting 282
shutter speed
  impact 157
simplicity
  avoiding visual noise 20
  in drawings 22
  in images in general 21
  in photos 22
  in screenshots 23
  in videos (basics) 23
  in videos (techniques) 257
single source publishing
  images 123
  videos 274
slow motion
  use in videos 262
software
  images of 167
speed
  of videos 305
stop motion
  tips for shooting 295
  use in videos 262
storyboard
  design 236
  need for 235
symbols
  designing 31
  file format 119
  finding 29
  pros and cons 28
  using 28

T

text
  pros and cons 13
  within images formatting 95
  within images using 33
  within images writing 37
  within videos using 33
  within videos vs. closed captions and voice over 240
  within videos writing 37
text callouts
  formatting 85
  pros and cons 82
thumbnails
  using in online documentation 65
time
  indicating in images 149
time lapse
  tips for shooting 295
  use in videos 262
timing
  in videos 305
titles
  figure titles 101
  for videos 233
translation
  cultural differences with visuals 41
  of images 125
  of screenshots 214
  of videos 301
trimming
  photos 165

U

UI
  avoiding confusion with in screenshots 193
updates
  facilitating of images 121
  facilitating of videos 303
user interface
  avoiding confusion with in screenshots 193
  showing 167

334

V

videos
  ambient noise 255
  authenticity 280
  backdrop in general 155
  backdrop with presenter 290
  closed captions 240
  design 219
  effects 262
  facilitating updates 303
  hypervideo 316
  interactive 316
  length in total 224
  length of modules 274
  lighting 286
  linking 268
  localizing 301
  music 253
  navigation 226
  placement 230
  player controls 228
  poster image 233
  preparing shooting 282
  presenter clothing 284
  presenter showing 245
  production 279
  pros and cons 15
  real vs. animation 222
  recording tips 292
  replacement for print 17
  simplicity 257
  speed 305
  standardization 276
  text including 33
  text vs. closed captions vs. voice-over 240
  text writing 37
  title 233
  translating 301
  types 220
  voice selection 251
  voice-over vs. text 240
  warnings 243
  ways of keeping simple (basics) 23
visual noise
  avoiding 20
visualizing
  rules 11
voice
  choosing 251
  tips for recording 298
voice-over
  use in videos 240
  voice choosing 251

W

warnings
  in videos 243
white balance
  with photos 162
whiteboard animation
  use in videos 262
window size
  in screenshots 186

X

X-ray image
  using 143

Z

zoom
  zoom and pan effect 262
zooming
  in images 79

335

More books by the same author:

**Technical Documentation Best Practices:
Planning and Structuring Helpful User Assistance**

Contents, Structure, User Navigation

**Technical Documentation Best Practices:
Visually Designing Modern Help Systems and Manuals**

Layout, Formatting, Templates

**Technical Documentation Best Practices:
Writing Clear and Helpful User Assistance**

Writing Rules, Tips, FAQ

Technical Documentation Short and Sweet

The Best Best Practices for Creating Clear and Useful Manuals, Help Systems, and Other Forms of User Assistance

Translating Technical Documentation Successfully

Things That You Should Preserve When Translating User Manuals and Online Help Systems (for translators)

For detailed information on all editions and on the available formats, visit *www.indoition.com*

indoition Starter Template

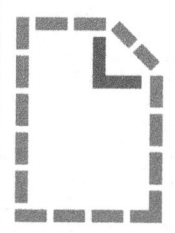

Professional technical documentation template

Many authoring tools don't come with a suitable template for creating clear, appealing user manuals, and setting up your own template from scratch can be time-consuming. The indoition Starter Template speeds up this task and prevents you from making costly strategic mistakes. It provides:

- a design that pleases the eye *and* communicates your message clearly
- paragraph styles and character styles that are efficient to use when writing and updating your documents

The Starter Template has been designed for Microsoft Word, OpenOffice, and LibreOffice, for A4 and Letter paper sizes. If you use a different paper size, basically all you need to do is to change the page margin settings.

Many other authoring tools can import Microsoft Word files (*.docx) and OpenOffice / LibreOffice OpenDocument Text files (*.odt) as well.

Key features:

- **no bells and whistles**—the template contains only what you and your users really need
- **automated styles** that eliminate a lot of manual formatting; optimized settings for automatic line breaks and page breaks
- **uses a time-tested**, systematic scheme for style names and keyboard shortcuts
- **works with all language versions of Microsoft Word, OpenOffice, and LibreOffice**—no need to edit style names and field codes if you're using a localized version of your authoring tool
- **well-prepared for being able to create online help from your document files as well** with the help of an appropriate single source publishing tool or converter
- includes **detailed instructions** on how to use the styles, and on how to change them if necessary

For detailed information, visit *www.indoition.com*.

CPSIA information can be obtained
at www.ICGtesting.com
Printed in the USA
LVHW010335221220
674784LV00015B/1056